SPACE, OBJECTS, MINDS, AND BRAINS

ESSAYS IN COGNITIVE PSYCHOLOGY

Essays in Cognitive Psychology is designed to meet the need for rapid publication of brief volumes in cognitive psychology. Primary topics include perception, movement and action, attention, memory, mental representation, language, and problem solving. Futhermore, the series seeks to define cognitive psychology in its broadest sense, encompassing all topics either informed by, or informing, the study of mental processes. As such, it covers a wide range of subjects including computational approaches to cognition, cognitive neuroscience, social cognition, and cognitive development, as well as areas more traditionally defined as cognitive psychology. Each volume in the series will make a conceptual contribution to the topic by reviewing and synthesizing the existing research literature, by advancing theory in the area, or by some combination of these missions. The principal aim is that authors will provide an overview of their own highly successful research program in an area. It is also expected that volumes will, to some extent, include an assessment of current knowledge and identification of possible future trends in research. Each book will be a self-contained unit supplying the reader with a well-structured review of the work described and evaluated.

Titles in preparation
Brown, *The Deja Vu Experience*
Gallo, *Associative Illusions of Memory*
Gernsbacher, *Suppression and Enhancement in Language Comprehension*
McNamara, *Semantic Priming*
Park, *Cognition and Aging*
Cowan, *Limits to Working Memory Capacity*
Coventry and Garrod, *Seeing, Saying, and Acting*

Recently published
Robertson, *Space, Objects, Minds, and Brains*
Cornoldi & Vecchi, *Visuo-spatial Representation: An Individual Differences Approach*
Sternberg et al., *The Creativity Conundrum: A Propulsion Model of Kinds of Creative Contributions*
Poletiek, *Hypothesis Testing Behaviour*
Garnham, *Mental Models and the Interpretations of Anaphora*
Engelkamp, *Memory for Actions*

For continually updated information about the *Essays in Cognitive Psychology* series, please visit www.psypress.com/essays

SPACE, OBJECTS, MINDS, AND BRAINS

Lynn C. Robertson

Routledge
Taylor & Francis Group
LONDON AND NEW YORK

First published 2004 by Psychology Press

Published 2022 by Routledge
2 Park Square, Milton Park, Abingdon, Oxon OX14 4RN
605 Third Avenue, New York, NY 10017

Routledge is an imprint of the Taylor & Francis Group, an informa business

Library of Congress Cataloging-in-Publication Data

Robertson, Lynn C.
Space, objects, minds, and brains / by Lynn C. Robertson.— 1st ed.
p. cm. — (Essays in cognitive psychology)
Includes index.
ISBN 1-84169-042-2 (hardcover)
1. Space perception. 2. Perception, Disorders of. I. Title. II. Series.
QP491.R585 2003
153.7'52--dc21

2003009120

ISBN 978-0-415-65189-9 (pbk)

To RM and his family,
and to all the patients who have willingly given their time and efforts
for the advancement of scientific knowledge despite the struggles
of their everyday lives.

CONTENTS

PREFACE

When I began studying spatial deficits in the early 1980s, I was amazed at the different ways in which perception could break down upon damage occurring to different areas of the human brain. Of course, neuropsychologists and neurologists had been observing the sometimes bizarre cognitive deficits that brain injury could produce for over a century, and many had developed practical bedside and paper-and-pencil tests to evaluate what types of spatial disorders were present. Remarkably, these tests were nearly 70% accurate in isolating the location of damage, a critical contribution to medical care before the invention of imaging techniques.

For the most part, cognitive psychologists who studied sensation and perception had never heard of the myriad of ways that perception could be altered by brain insult and were unaware of the rich phenomenon that would eventually prove invaluable to scientific efforts to understand perception and attention in addition to the neural mechanisms involved. In those early days, "cognitive neuroscience" was a new area of study that Mike Gazzaniga and Mike Posner, with funding from the James S. McDonnell Foundation had begun to introduce to the scientific community, but it was often met with either resistance or ennui from an academy that had divided into separate turfs.

I sat on one of those turfs until I discovered my myopia when I took a position at the Veterans Administration medical facility in Martinez, California as a research associate for Pierre Divenyi. There I was introduced to a neurology ward, and my eyes were rapidly opened to the fertile ground on which I had landed. I immediately started learning everything I could about the types of cognitive problems that occurred after damage to different areas of the human brain. I was especially struck by spatial deficits that resulted in parts of visually presented displays disappearing from conscious awareness, as if they did not exist at all. Other patients remained conscious of the items in a display, but the perception of their spatial locations was drastically altered.

I quickly changed my experimental approach from a model based on an isolated scientist doggedly pursuing the answer to a specific problem in her laboratory to one that embraced cross-disciplinary collaboration and an appreciation for scientific diversity. The patients themselves became much more than "subjects" or "participants." They were individuals struggling with their problems every moment of every day. I discovered that visual deficits were far more restrictive and problematic than I ever thought possible, and rehabilitation measures for some problems were practically nonexistent. I discovered that neurological patients presenting with unilateral neglect were more likely than any other stroke groups to end up in nursing homes in the long run. Visual-spatial disorders became more than a scientific interest for me. The translational value of my work came into view, and understanding visual-spatial processing from both a cognitive and neuroscience point of view became a lifetime goal.

This book represents what came of that goal. It would never have been written if Henry Roediger had not suggested my name to Alison Mudditt, then the publishing director at Psychology Press, as someone who might contribute to the new Psychology Press series, Essays in Cognitive Psychology. Alison's replacement, Paul Dukes, deserves special credit for picking up where she left off and taking the manuscript through to press. Also, the Veterans Administration Medical Research Council, the National Science Foundation, and the National Institutes of Health receive my special thanks for supporting my research over many years.

I had not been thinking about writing a book when I was approached by Alison a few years ago, but since I was told it could be a monograph centered on my own work, the task seemed easy, and I thought it might be fun. I expected to have a draft done the following summer. Four years later, I am still wondering if I have the story right, but there must be an end to writing such a book, and that end has come. I have learned a great deal more than I expected along the way, and during this time the study of space and objects has evolved within cognitive neuroscience in ways that I find encouraging. I am sure I have left several important bits of information out, and I apologize to those who have been omitted, but again, one must stop somewhere.

Writing this book also gave me the opportunity to think more deeply about how the different aspects of my work fit together and how to communicate the sometimes controversial, if not idiosyncratic, positions that I have taken. I hope I have succeeded if in nothing else, to stimulate ideas and debate in some small way.

Critically, without the collaboration and encouragement from my colleagues, this book would have never been written. I cannot thank enough my long-time colleagues Robert Knight and Robert Rafal for teaching me the finer points of behavioral neurology and neuropsychology. They welcomed me to accompany them on their ward rounds and into their clinics, patiently explaining neurological causes, treatments, and probable outcomes of various disorders. They were willing to answer my naïve questions without laughing (well, maybe sometimes) and made me appreciate the art of medical diagnosis and the clinical decision-making process. They demonstrated how to accommodate to a patient's deficits and to select bedside tests wisely when confronted with patients who were often fatigued, confused, distracted, or in pain. Their respect for their patients was contagious and rekindled my desire for the humanistic side of behavioral science.

I am also very grateful to my colleagues Anne Treisman, Steve Palmer, Richard Ivry, and Dell Rhodes, who were influential in the theoretical developments that led to this book as well as in some of the experiments and interpretations that form the basis of selected arguments. I savor the many good meals with these individuals and the interesting conversations.

The hours of testing the patient, RM, and discussing results with Anne Treisman were a complete delight, and her reading every word of an earlier draft of this book has surely increased its scholarship. She has been a friend and mentor for many years, and I feel privileged to be continuously challenged by her probing questions and thoughtful comments. I am also greatly indebted to Krista Schendel and Alexandra List, who read an earlier draft of many of the chapters and contributed substantially to the final product. None of these individuals should be held responsible for my misinterpretations or mistakes, but each has provided valuable comments and insights.

I also wish to thank many current and former students who worked long hours on individual studies that molded my thinking; studies that are referenced at different points throughout this book. These individuals include Lisa Barnes, Lori Bernstein, Shai Danziger, Mirjam Eglin, Robert Egly, Stacia Friedman-Hill, Marcia Grabowecky, Min-Shik Kim, Marvin Lamb, Ruth Salo, and Krista Schendel. Without their labor and fortitude, none of this would be possible. Several of my current students, Joseph Brooks, Mike Esterman, Alexandra List, and Noam Sagiv, will surely contribute to the future understanding of the topics covered in this book, given the projects they are working on at the present time. A special thanks goes to Ting Wong, who helped prepare the manuscript and the many figures, and to my colleague Jack Fahy, who has become an integral part of the neglect research.

Last but not least, I owe much to my partner in life, Brent Robertson. His patience and support are the most important contributions to the writing of this book, and he has encouraged me all along the way.

1

CHAPTER

Losing Space

Where is the Triple Rock cafe?
    It's that way.
How far is it?
    About a mile after the next traffic light.
Is it on the right or left?
    It depends which way you're walking.
Is it further than Jupiter's?
    Yes, especially if you stop for a brew.

—As heard on a Berkeley street corner (or could have been)

We all ask these kinds of questions to get where we want to go. Each landmark we use (the pub, the streetlight) is different, but the space we travel through seems much the same—just a void between destinations. We refer to space as "cluttered" when it becomes overly filled, and we look through space as if it is just air between one object and another.

Yet space is also a thing, and regarding perception, it is a special kind of thing. Unlike outer space, perceptual space is not infinite. It has boundaries. When we look upward toward the sky, space has an end. It stops with the day's blue sky or the night's black background behind the moon and stars. Space is not a void in our mind's eye. Its depth, volume, and boundaries are all part of the brain's creations given to us in perceptual awareness. Just like objects, spaces have form and can be conceptually and physically different. The space inside a tennis ball is different from

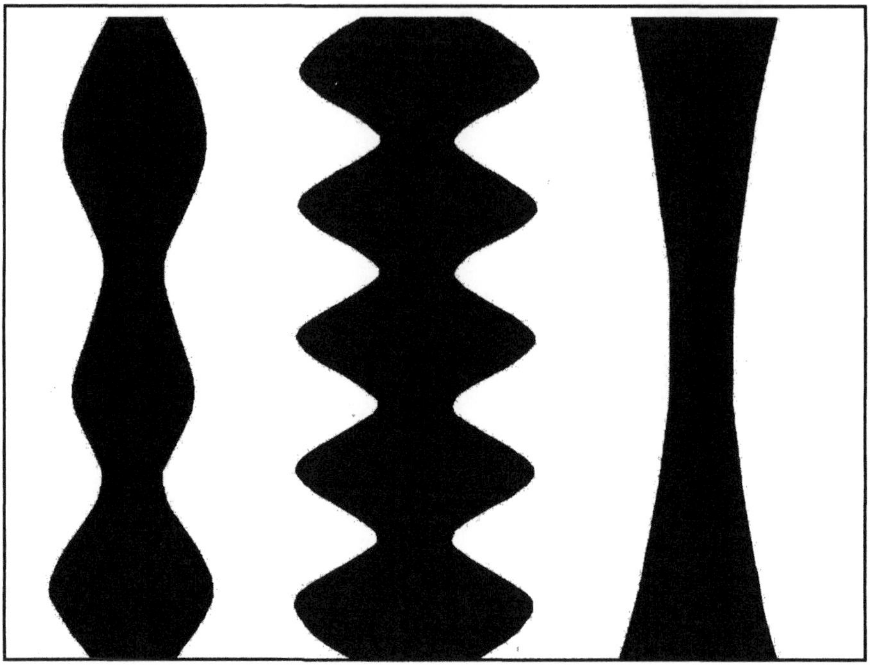

FIGURE 1.1. Example of a figure in which the black portions are more likely to appear as figure and the white portions as ground. The ground appears as a single space unless attention is directed to it.

the space between the sun and the earth. The space between atoms is different from the space between houses. The spaces between a group of boundaries (Figure 1.1) have a form all their own, although we perceive them as a unified space behind foreground objects.

Perceptual space, unlike physical space, can be changed by the perceiver. When attention is directed to the space between the boundaries of Figure 1.1, that space changes from being part of a unified background to a set of unique forms themselves. When a few lines are connected on a blank sheet of paper, they create a space within the boundary of what we see as a square and another space outside the boundary of what we see as a sheet of paper. A few mere lines can change one space into two (Figure 1.2). More lines still (Figure 1.3) can change two spaces into three. We typically call these spaces objects or shapes (diamond, square, sheet of paper) and often ask questions about how the configuration of lines (as well as shadows, colors, contour, etc.) contributes to *object* perception. Alternatively, we might ask how the configuration of objects changes perceived space. It turns out that objects can change one perceptual space into many, and the configuration of lines can shape space, changing its scale or volume.

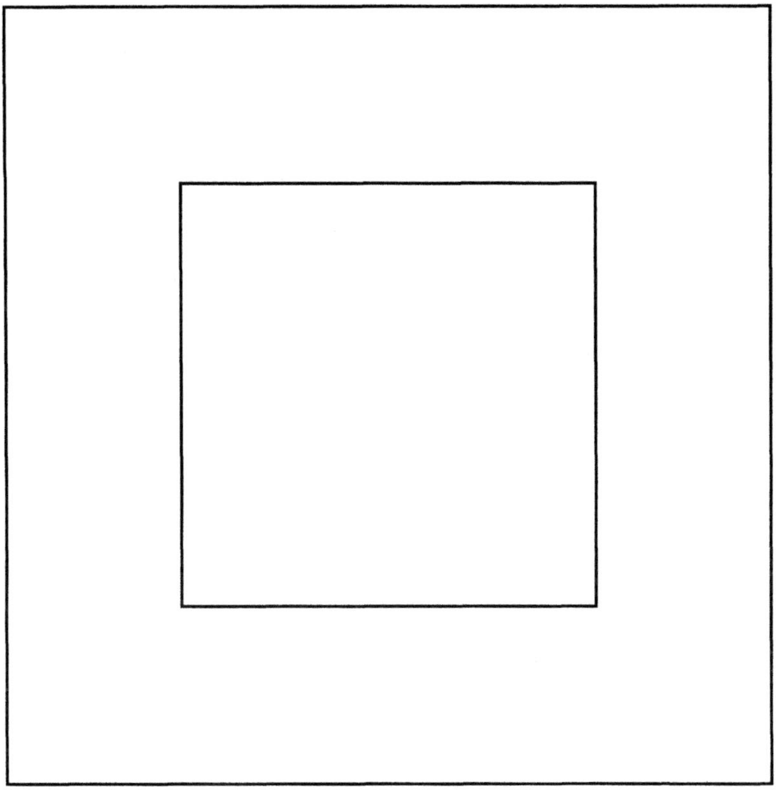

FIGURE 1.2. The smaller square defines one space and the larger square another.

It is not difficult to see how readily this leads to a scientific conundrum. If space defines objects, then we need to know how space is represented to know when or how an object will be perceived. But if objects define space, then we need to know how objects are represented to know how space will be perceived. After a century of psychological research, we know only a little about either and even less about how the two interact to form the world we see.

It has been customary in much of cognitive research to assume that space is constant, with objects defined by the contours drawn over this space. After all, we move from one item to another through a single, metric three-dimensional (3-D) outer space, and when we scan a visual scene, attention seems to move in the same way. But we tend to forget that perceived space, as well as all the spaces within the boundaries we call objects, is malleable. The space outside our skin, for all practical purposes, may be constant, but perceived space is not. It can explode and break into pieces or disappear altogether. This fact becomes painfully

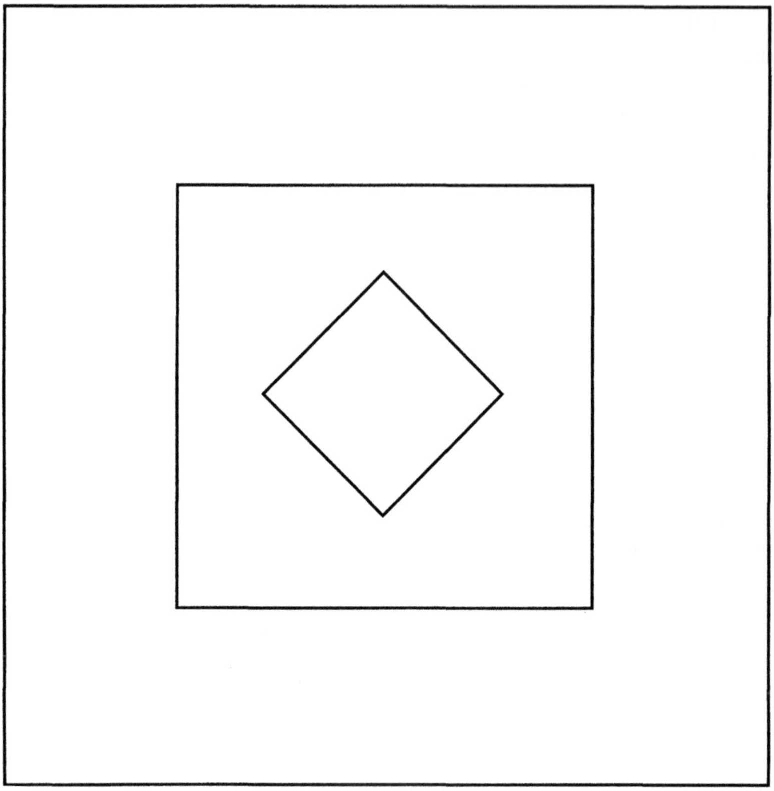

FIGURE 1.3. Adding a diamond to Figure 1.2 creates an additional level, now with three tiers.

obvious when the brain goes awry and space perception breaks down. The ways this can happen and why it happens in some ways but not others form the basis of what is to follow.

☐ When There Is No "There" There (Balints Syndrome)

Imagine yourself sitting in a room at the Chicago Art Institute contemplating Caillebotte's *Paris Street: Rainy Day* (Figure 1.4). You admire the layout of the buildings as well as the violations the painter has made in proportion and symmetry. The play of water and its reflection off the stones catches your eye, and then your attention might be drawn to the pearl earring of the woman in the foreground. It looks delicate and bright against the darkness of that part of the painting. You may even wish you

FIGURE 1.4. Caillebotte's painting *Paris Street: Rainy Day*. (Gustave Caillebotte, French, 1848–1894, *Paris Street: Rainy Day*, 1877, oil on canvas, 212.2 × 276.2 cm, Charles H. and Mary F. S. Worcester Collection, 1964.336. Copyright © The Art Institute of Chicago. Reprinted with permission.)

were part of the couple walking arm in arm down a Paris street under a shared umbrella.

Now imagine you look again. There is only an umbrella. You see nothing else. Your eyes are fixed straight ahead of you, yet that umbrella seems to fill your whole visual world. But then, all of a sudden, it is replaced by one of the cobblestones. You only see one. Are there others? This image might stay with you for what seems like minutes, but then, without notice, the cobblestone disappears and is replaced by a single gentleman. Next, the pearl earring may take over. It looks like a white dot of some sort. For you it does not look like an earring, since it attaches itself to nothing. You don't even know where it is. Is it to your left or right? Is it far or near? Is it closer to the floor or the ceiling? Sometimes it looks very small, other times, very large. It may change colors from white to sienna to bluegray (other colors in the painting). Since you don't know where it is, you cannot point to it, and if it were a real pearl hanging in front of you that you wanted to hold, you would have to make random arm movements until you touched it by chance. Once in your hand, you could readily identify it as a pearl earring and you could put it on your own ear easily (you have not lost motor control or the spatial knowledge of your own body). The space "out there," whether the spatial relationship between one object and another or the spatial relationship between a part of you and the object you see, is no longer available. Somehow your brain is not computing those spaces. There is no there there.

This is a scenario that fortunately happens only rarely and is known to neurologists and neuropsychologists as Balints syndrome. The syndrome can vary in severity and recovery from it is erratic. In the few "pure" cases reported in the literature, there was damage to both sides of the posterior portions of the cortex without the loss of primary visual or motor abilities or other cognitive functions (e.g., language). The syndrome has been noted in a subset of dementias (Hof, Bouras, Constantinidis, & Morrison, 1989), but it is often difficult to sort out which deficits are due to the dementia per se and which are due to a loss of spatial knowledge in these cases.

The loss of spatial knowledge with bilateral posterior brain damage was first reported in 1909 by the neurologist Rezso Balint in a patient with lesions in both hemispheres, centered in the occipital-parietal lobes (Figure 1.5). The deficits that occur when these areas are damaged on both sides of the brain were later confirmed by Holmes and Horax (1919) and Holmes (1919) who reported a number of additional cases of the syndrome. The clinical syndrome is defined by three main deficits: (a) simultanagnosia, or the inability to see more than one object at a time, (b) optic ataxia, or the inability to reach in the proper direction for the perceived object, and (c) optic apraxia, or a fixation of gaze without primary eye movement deficits (what Balint called "pseudoparalysis of gaze").

Some of the questions about normal perceptual processing that these cases bring forth are as follows:

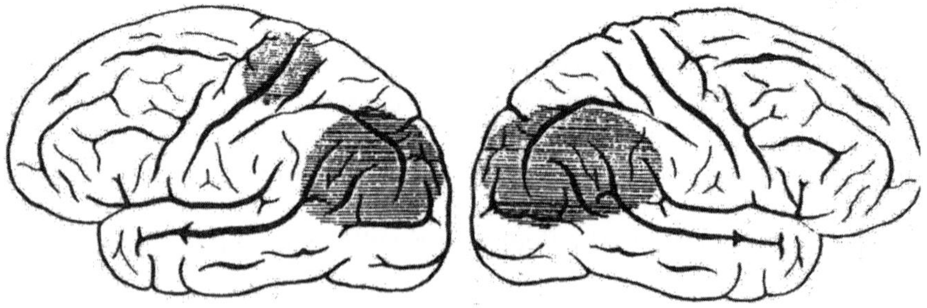

FIGURE 1.5. Areas of "softening" in a Balints patients upon postmortem examination. (From "Seelenlähmung des 'Schauens'," optische Ataxie, räumliche Störung der Aufmerksamkeit by Rudolph Bálint. Copyright © 1909. In the public domain.)

1. If space is represented as a single property or feature, how can body space be preserved while space outside the body is disturbed?
2. How can even a single object be perceived without a spatial map?
3. What are the characteristics of the single object that enters awareness when perceptual space disappears?
4. Why would a spatial deficit result in the misperception of an object's color?

These questions and more will be addressed in the chapters that follow, and the answers (as preliminary as some may be) have revealed many interesting aspects about how brains bind together information in our visual worlds and the role that perceptual space plays in this process. Space not only tells us where things are but also helps us see what they are.

☐ When Only Half Is There (Unilateral Neglect)

Consider again Caillebotte's painting reprinted in Figure 1.4. This time you first see the edge on the right with the foreground figure of part of the back of a man. After this you might see a portion of the woman holding the umbrella, but then all you might see is the right edge of the woman and the umbrella along with the earring the woman is wearing. Each bit that comes into view extends toward the ceiling and floor and you look up and down to see buildings in the background (perhaps deviating somewhat between upper and lower parts). At some point you stop scanning leftward, perhaps seeing only the half of the painting that extends from somewhere in the middle to the rightmost edge. You see the couple walking arm in arm and in the center of the painting that you see, although only the right half of each might be visible to you. You might even admire the painting's beauty and proportion, but you have missed the left side of space as well as the left space of objects within the right side of the painting that remains visible to you.

If you were familiar with Caillebotte's painting, you might wonder where the left side went. Did some vandal destroy it? If you were not familiar with the painting, you would not know that the triangular building that juts out toward a normal viewer on the left side is even there. It is as if half of space has disappeared, but since you are not aware of it, you think that the space you still see is complete.

This type of perceptual deficit, known as hemineglect or unilateral visual neglect, is produced by damage to areas on one side of the brain (usually the right) and is generally associated with damage to parietal lobes (although frontal and subcortical neglect have also been observed). The neglect syndrome has become familiar to most psychologists who study visual cognition, although it was unknown to a majority before the emergence of cognitive neuroscience. The cortical damage that produces hemineglect is limited to one hemisphere of the human brain and often (but not always) includes some of the same areas that produce Balint's syndrome through bilateral damage. When damage is isolated to one side, space contralateral to the lesion (contralesional) seems to disappear.

Hemineglect is much better understood today as a result of increased interest in the syndrome, new techniques to study the human brain, and the development of new behavioral tests to understand the cognitive and neural mechanisms involved. For instance, it seems to be linked to spatial attention in predictable ways. When items are present on the right side (e.g., the man's back, the woman, the earring), attention seems to be attracted there and become "stuck," either preventing or delaying attending to items on the left of the current focus (Posner, Walker, Friedrich, & Rafal, 1984). The magnitude of neglect (i.e., the time it takes to notice something on the left side) can vary with the attentional demands of information on the right side (Eglin, Robertson, & Knight, 1989; Eglin, Robertson, Knight, & Brugger, 1994). Neglect can have motor and/or perceptual components depending on the area of the brain affected (see Bisiach & Vallar, 2000), and it can be both space-based and object-based (see Berhmann, 2000). For instance, the left side of the umbrella in Caillebotte's painting might be neglected, or the left side of the lady in the couple.

Drawings by patients with neglect reveal this pattern better than my discussion (Figure 1.6). Note that the patient drawings shown in Figure 1.6a include the right side of different objects across the scene but omit those on the left side. The left side of the house, the window on the left of the house, the left side of the tree to the left of the house, and the left side of the fence can all be missing. The patient drawings in Figure 1.6b show that the right side of the cat was sketched with distinguishing details like the tail included, but the left side of the cat in Figure 1.6b was poorly drawn and the tail was left out completely. If an artist with neglect were asked to copy Caillebotte's painting, the left side of the umbrella might be missing, as might the male partner of the strolling couple (he being to the left side of the woman) as well as the left side of the painting itself. The drawing might appear something like the cartoon creation shown in Figure 1.7.

Object- Versus Space-Based Attention: Is There a Dichotomy?

The observation of neglect for objects as well as space has been used to support arguments for separate object- and space-based modes of atten-

FIGURE 1.6a. Examples of drawings by three patients with left visual neglect showing neglect on the left side of objects. (Reprinted by permission from Gianotti, Messerli, & Tissot, 1972, with permission of Oxford University Press.)

FIGURE 1.6b. Examples of drawings by a patient with left visual neglect showing neglect of the left side of a cat drawn from a standard (top) depicting either the left or the right sides. (Reprinted by permission from Driver & Halligan, 1991, with permission of Psychology Press, Ltd., Hove, United Kingdom.)

FIGURE 1.7. Cartoon rendition of what the drawing of a patient with neglect might look like if asked to recreate the painting in Figure 1.4.

tion. In behavioral studies with unimpaired individuals, it is very difficult to separate the two. Objects inhabit space, and when attention is directed to an object, it is also directed to the space it occupies. Reports of object- vs. space-based neglect have been used to support the dichotomy of two separate attentional systems, one directed to objects and one directed to space.

As intuitively appealing as it might be to apply the neuropsychological evidence as support for two modes of attention, object-based neglect is not easy to specify objectively. What does it mean to leave out the left side of the strolling couple? Is this a case of object-based or space-based neglect? If the couple were considered as two separate people (i.e., two perceptual units), then this would appear to be a case of space-based neglect. The one on the right is seen, while the one on the left is not. But if the couple is considered as one pair (i.e., one perceptual unit), then the same errors might be viewed as a case of object-based neglect. The half on the right side of the pair is perceived, while the half on the left is not. Consistently, the picture as a whole can be thought of as one perceptual unit. If a patient with neglect drew the left side of the painting but not the right, this would be considered space-based neglect. But this too could be a case of object-based neglect. The left side of the picture or object is missing. One can see how arbitrary all this can be. Almost any pattern of neglect can be used as an example of either object-based or space-based neglect depending on the frame of reference adopted by the observer (examiner or scientist). The question then becomes, What frame of reference is the patient using?

The examples I've described to make this point have used a famous painting, and the drawing in Figure 1.7 is completely fabricated. However, there are published drawings from patients with hemineglect that demonstrate the same point (see Halligan & Marshall, 1997). Perhaps the most well known are those of an artist with neglect who drew a self-portrait at different stages of recovery (Figure 1.8). Note that in all the drawings the left side of the face and of the picture is either missing entirely or at least more disturbed than the right. In the first drawing the left eye is missing, but in later drawings it is present. If we consider eyes as a pair, then the first drawing would be an example of object-based neglect, but if we consider each eye as an individual unit, then this would be an example of space-based neglect.

The foregoing discussion has not simply been an exercise in establishing how complex the neglect phenomenon can be. It has important implications for how we consider normal visual cognition and the frames of reference that define the visual structure of the perceived world. It should be clear from these few examples that the terms *object-based* and *space-based* are slippery concepts, and this is also the case when thinking about normal vision. It depends on what the interpreter calls an object and what space is selected to provide the frame of reference.

This problem will become especially relevant when the issue is explored more fully in Chapter 4. It is also relevant for neurobiological arguments that object- and space-based neglect can be linked to separate cortical streams of processing (see Figure 1.9), a dorsal pathway that functions to reveal *where* things are and a ventral pathway that processes *what* things are (Ungerleider & Mishkin, 1982). More recently, it has also been extended to functional hemispheric difference (Egly, Rafal, Driver, & Starreveld, 1994). The left hemisphere is said to be more object-based, while the right hemisphere is argued to be more space-based.

It should be quite obvious by now that objects and space are not nearly as easy to dissociate as the concepts themselves imply. It follows that attributing them to dissociable neural systems is problematic for the same reason, and the arguments for doing so have in many cases been entirely circular. Without a good understanding of how the visual system defines an *object*, how can we know when hemineglect is due to neural mecha-

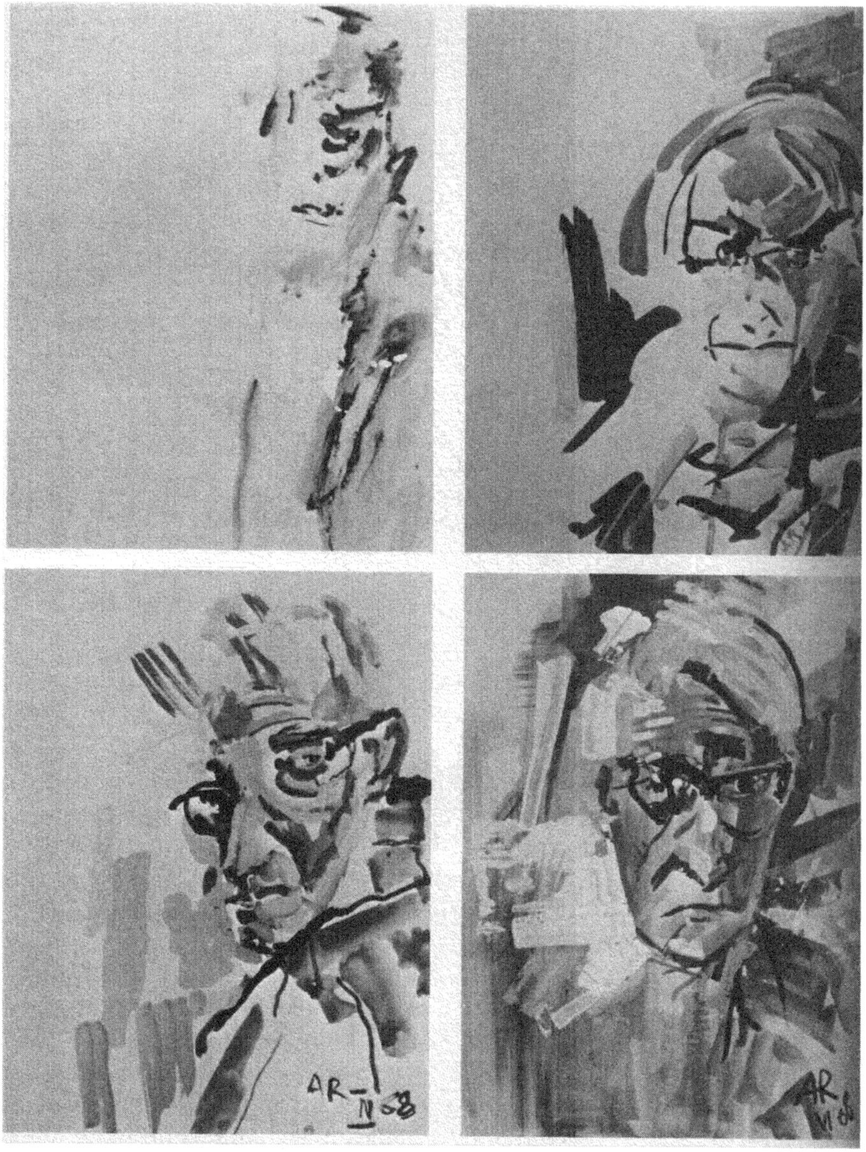

FIGURE 1.8. Self-portrait by an artist who suffered a stroke, causing left neglect. The drawings are at different stages of recovery starting with the upper left. (Copyright © 2003 Artists Rights Society (ARS), New York/VG Bild-Kunst, Bonn. Reprinted with permission.)

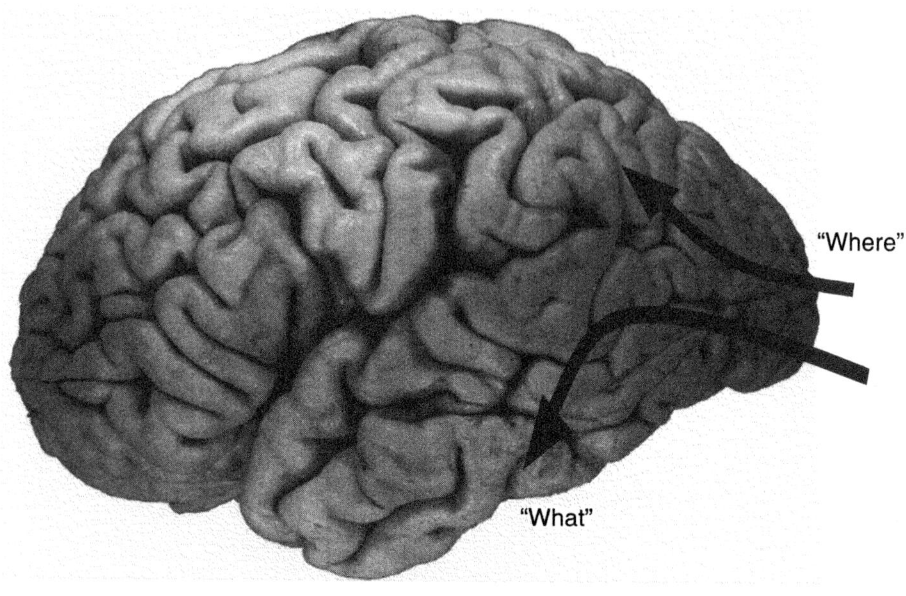

FIGURE 1.9. Drawing showing two major visual processing pathways through the cortex: A dorsal pathway that is said to process "where" or "how," and a ventral pathway that is said to process "what."

nisms that are object-based? Likewise, without a good understanding of how vision defines *space*, how can we know when hemineglect is due to neural mechanisms that are space-based?

In the chapters that follow, I will argue that the space vs. object dichotomy should be thought of instead as levels in a space/object hierarchy of reference systems. There are objects within objects within objects that contain spaces within spaces within spaces (Figure 1.10). Another way of describing this relationship is as a system of hierarchically arranged spatial coordinates. In Figure 1.10 there are lines that demarcate the borders of enclosed spaces that we call squares or boxes with larger lines that demarcate the borders of another space that surround the first, and so on and so forth. Box 3 represents the smallest object, and coordinate 3 represents the space that defines it. Box 2 represents the next-to-the-largest object, and coordinate 2 represents the next-to-the-largest space that defines it. Box 1 is the most global object in the figure and is defined by the largest coordinate.

Within a system such as this, object-based neglect simply represents another case of space-based neglect but within different spatial coordinates. If the spatial coordinates of box 3 are selected, then spatial neglect

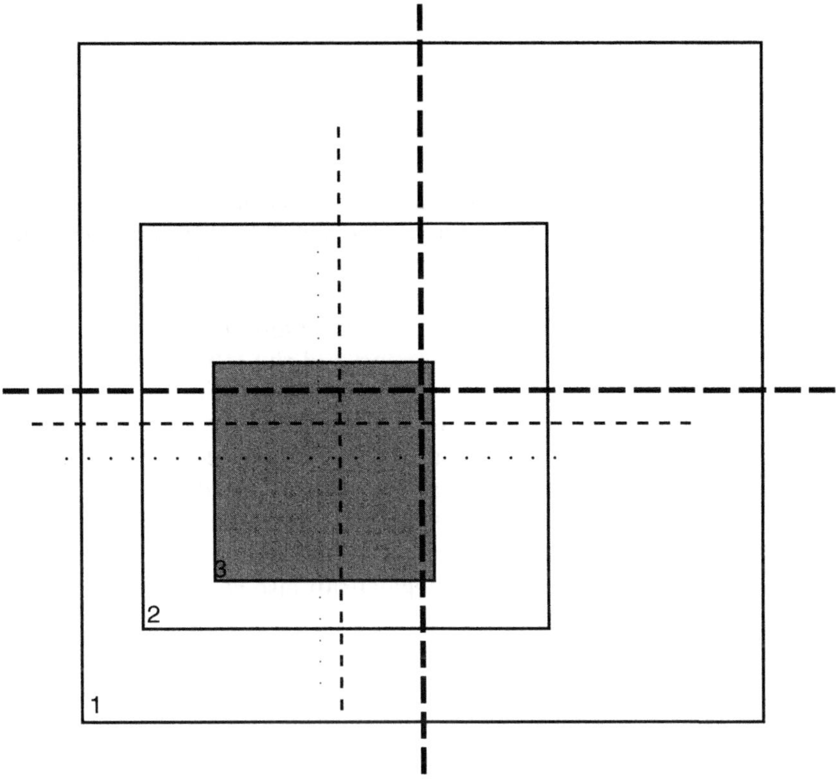

FIGURE 1.10. Hierarchically organized set of squares with the coordinates that define them centered on each. Square 1 is the most global level, and square 3, the most local.

will be manifested within the local object, and if the spatial coordinates of box 1 are selected, then spatial neglect will be manifested within the more global object. So if attention were drawn to the couple in Caillebotte's painting, neglect would be to the left of the vertical coordinate centered on the couple (the reference frame that defines the stimulation within it as on the left or right). If attention were drawn to the painting as a whole (a more global reference frame), neglect would be to the left of the coordinate centered on the painting. If attention were drawn to the umbrella (a more local reference frame), neglect would be to the left of the coordinate centered on the umbrella.[1].

Notice that in this account there are not two types of hemineglect (object- vs. space-based). Rather, hemineglect is neglect of the left side of whatever reference frames control the allocation of spatial attention at the moment (whether volitionally or automatically selected). To make this case even more concrete as well as clinically relevant, Figure 1.11

shows the performance of a patient with neglect tested in my laboratory who was asked to circle all the As in a display that extended across a full page (Figure 1.11a) and when the display was clustered into two vertical columns (Figure 1.11b). This patient would be classified as having object-based and space-based neglect. When the page is divided into columns, performance demonstrates awareness of the column on the left side of the page showing that the column (i.e., what is called the object in this case) was not neglected. More accurately, the spatial frame that defines left and right in *each* column was represented and the left side of the vertical axis of each was neglected. When the display was not clustered into columns, as in Figure 1.11a, the spatial reference frame that defines left and right was centered on the page and the left side of this larger frame was neglected.

This description does not negate the idea that neglect can be object-based. It is object-based to the extent that each "object" is defined by a spatial coordinate, with the vertical axes of that coordinate determining what is left and what is right. The difference is that object-based neglect is not a separate manifestation of the neglect phenomenon. Patients who show what is called object-based neglect can also show space-based neglect in the common parlance. But note that the same lesion can produce both, and it is the space within each object in this object/space hierarchy that is neglected. Evidence consistent with this explanation of neglect will be discussed in Chapter 3 in far more detail.

Before leaving this section, it should be noted that all of these problems in knowing when an object/space is treated like an object or like a portion of space can also be applied to normal perception. The world contains multiple objects at multiple spatial levels and in multiple spatial locations. If there are truly space-based and object-based attentional mechanisms, then the ways that perceptual systems determine what an object is and what a space is in a complex scene seem fundamental.

☐ Not There but There (Integrative Agnosia)

Suppose that instead of missing the left side of Caillebotte's painting, you perceived all the items within it but with different objects in different places. The umbrella might be seen at the top left, with the person in the foreground somewhere in the center. The gentleman with whom the woman is strolling might appear over to the right toward the top, and cobblestones might be scattered here and there. You aren't looking at a Picasso. The Picasso is in your mind. The computation of the spatial relationships between different objects in the painting has been disturbed, and you see only an unorganized display.

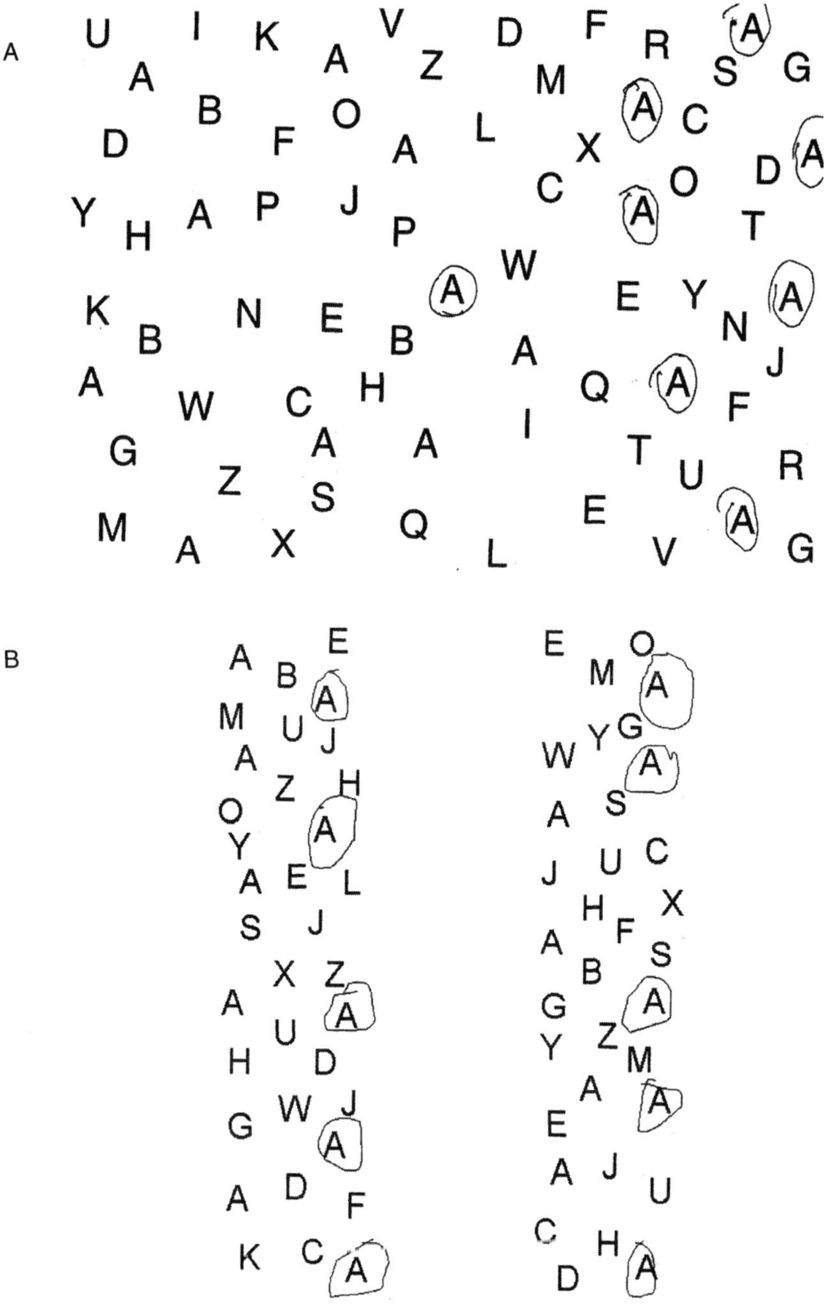

FIGURE 1.11. Examples from a patient with left neglect who showed both space-based (a) and object-based (b) neglect when aksed to circle all the As he could find in the displays.

Hemispheric Differences in Object/Space Perception

The drawings of patients with the type of deficit just described can be revealing. Figure 1.12 shows a reproduction (bottom right) of the drawing of a complex nonsense pattern (the Rey-Osterreith figure, shown at the top of the figure) by a patient with right hemisphere damage but without neglect. Notice that in the copy the details of the test pattern do not come apart in a totally random way. Features that are displaced appear as perceptual units or whole objects in themselves. The circle with dots in it remains a circle with dots in it. The track-like figure remains intact. Its details are not scattered in random fashion, as would be expected if the defining features of the objects, such as lines and angles, had also become spatially uncoupled. Another example is shown in Figure 1.13. For the patient with right hemisphere damage, the local objects are drawn correctly, while the global object is not. One could say that the

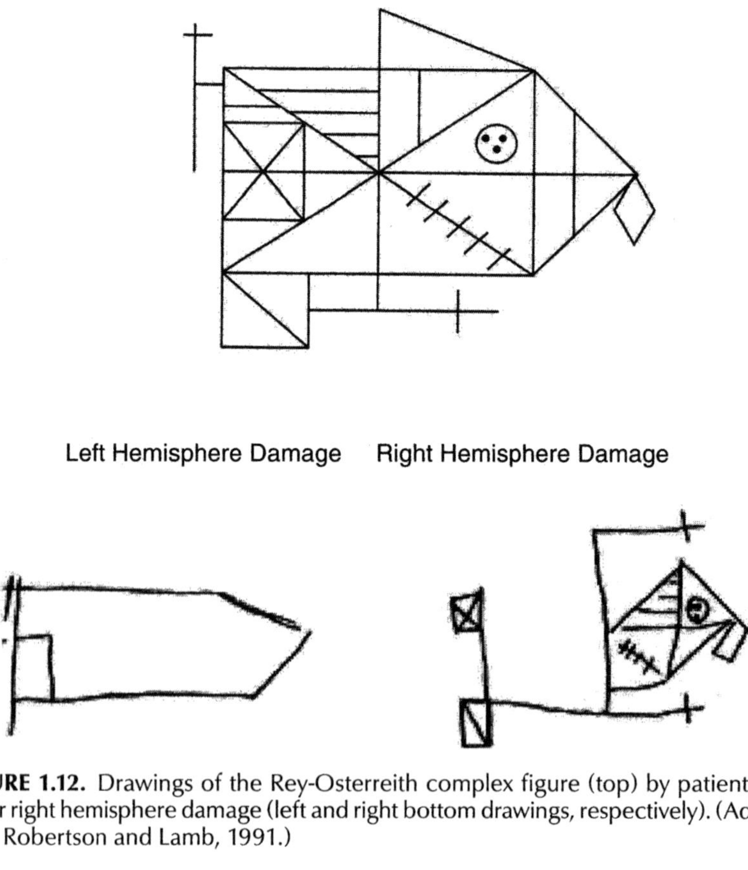

FIGURE 1.12. Drawings of the Rey-Osterreith complex figure (top) by patients with left or right hemisphere damage (left and right bottom drawings, respectively). (Adapted from Robertson and Lamb, 1991.)

drawing is of objects, but their spatial locations are not correct. Another way of saying the same thing is that local objects retain their spatial integrity, while global objects do not.

This type of problem is most often observed with lesions of the posterior right hemisphere that extend into ventral pathways. Consistently, functional imaging studies with normal perceivers have shown more right than left hemisphere activation when attending to global levels of a stimulus (Fink et al., 1996; Han et al., 2002; Heinze, Hinrichs, Scholz, Burchert, & Mangun, 1998; Mangun, Heinz, Scholz, & Hinrichs, 2000; Martinez et al., 1997; Yamaguchi, 2002) (see Figure 1.14, for one example). The exact locations that produce these effects is of some debate, the details of which will be touched upon in a later chapter. Let it suffice here to say that global processing and right hemisphere function has received a great deal of converging support.

Left hemisphere damage produces a complementary problem. Local objects are either missed or incomplete, while global objects remain relatively intact. Figure 1.12 (bottom left) shows a copy of the Rey-Osterreith complex pattern drawn by a patient with left hemisphere damage. The

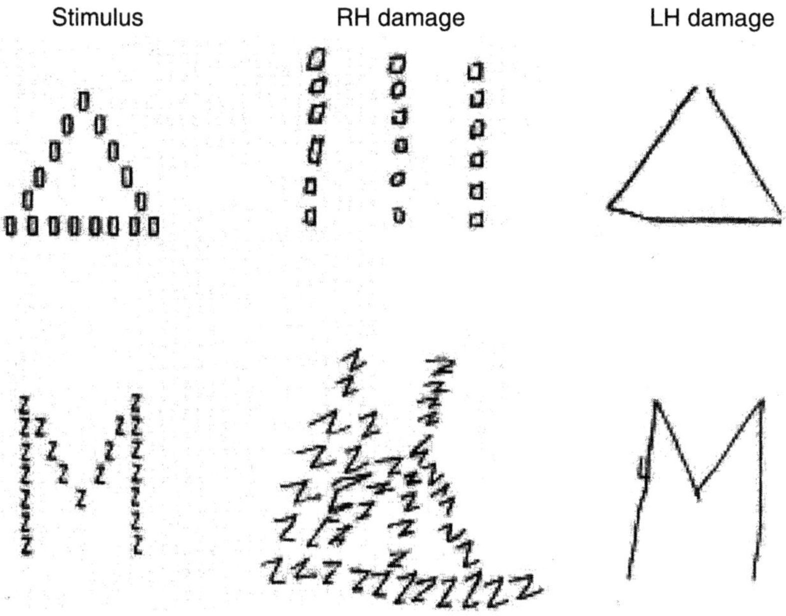

FIGURE 1.13. Examples of drawings of global letters and shapes created from local letters and shapes by patients with right (RH) or left hemisphere (LH) damage. (Adapted from Delis, Robertson, and Efron, 1986.)

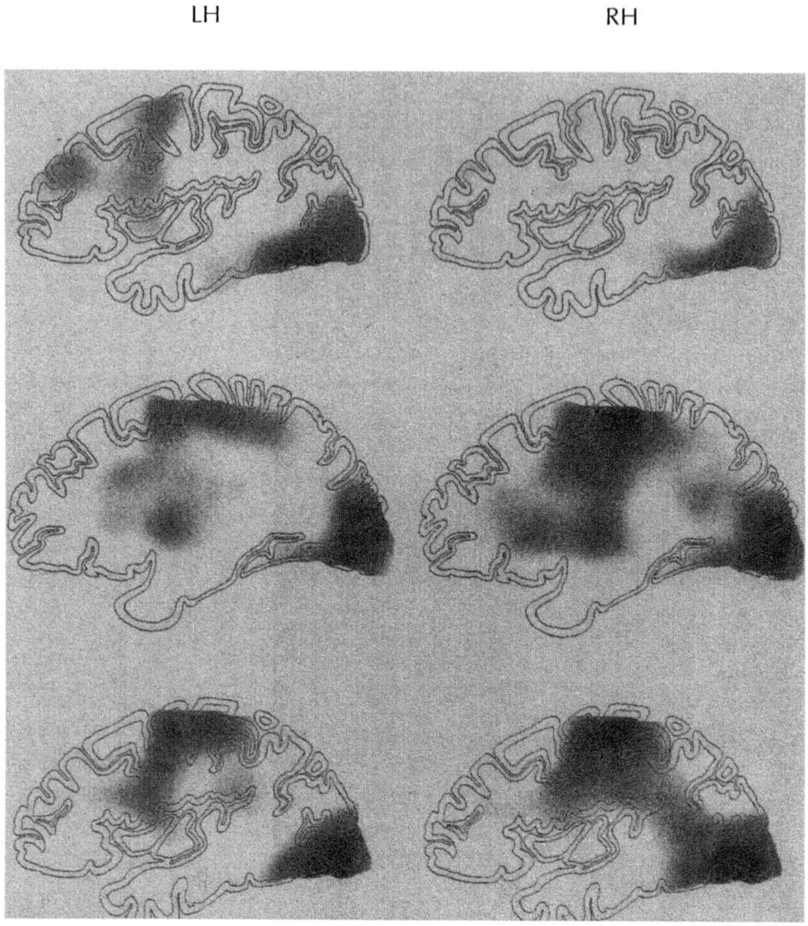

FIGURE 1.14. PET images showing more activation of the left hemisphere (LH) when attending to local information and more activation of the right hemisphere (RH) when attending to global. (Adapted from Heintz et al., 1998.)

global form is similar to the test pattern, but the local forms are sparsely represented or not at all (Figure 1.13). In Figure 1.13, the global M and triangle are correct, while the local L is not, and local rectangles are absent.

These deficits have been observed in groups of patients with damage centered in the left hemisphere (Robertson, Lamb, & Knight 1988). Again, imaging data have confirmed the hemispheric asymmetry of these perceptual differences and their relationship to the left hemisphere (Figure 1.14). When normal individuals attend to local elements, there is more activation in the left hemisphere than in the right (Fink et al., 1996; Han et al., 2002; Heintz et al., 1998; Martinez et al., 1997; Yamaguchi, 2000).

I will not discuss these deficits to any great extent in the chapters that follow, as Richard Ivry and I have done so at length under a separate cover (Ivry & Robertson, 1998). But there are several points from the study of global and local differences that may help put object and spatial deficits in context. First, there is the need to think of hierarchical relationships. Global and local levels in a stimulus are inherently relative, with a hierarchy of space/objects from higher level global to lower level local levels. Referring to Figure 1.10, again consider the most local box (box 3) and the most global (box 1) in that display. Patients with right hemisphere damage centered in posterior ventral areas would most likely have an altered representation of box 1, leaving the correctly perceived box 3 nowhere to go but into a wrong location. Patients with left hemisphere damage would have an altered representation of box 3, but because they would maintain the space/object perception of box 1, box 3 would be located correctly.

Ivry and Robertson (1998) argued that global and local deficits emerged from a problem in attending to *relative* spatial resolution (hypothesized as beginning with an asymmetry in attending to the relevant spatial frequency channels that provides certain basic visual features of perception) (see also Robertson & Ivry, 2000). Whether this theory turns out to be correct or not, it is clear that global/local (or part/whole) processing deficits are not the same type of spatial deficits as those observed when half of space disappears (hemineglect) or when all of space except for one object disappears (Balint's syndrome). However, the hierarchical organization of things in the external world must be taken into account in any theory of hemispheric differences based on these deficits. Given the different brain regions that contribute to different visual-spatial problems, it is not surprising that there would be differences in how space is utilized in object perception when damage occurs.

In sum, object and spatial deficits come in many guises, but may best be described in an object/space hierarchy. Although this conceptualization may seem like a small change, in fact, the types of questions that arise and the interpretation of data are clearly different. The question of how cognitive and neural mechanisms operate within each level of object/space and how that level is selected seems critical if we are to understand the relationship between representations of objects and space and how they are associated with brain function. In the following chapters, I will outline some of what we know about this relationship and venture into what it may mean for the very basis of conscious awareness itself.

2

CHAPTER

Object/Space Representation and Spatial Reference Frames

In Chapter 1, I argued for a hierarchical organization of spatial coordinates that define object/spaces at several levels in perception (akin to Rock's, 1990, proposal for a hierarchical organization of reference frames). But in order to think about how this object/space hierarchy could be useful for perception and attentional selection, we need to know what spatial properties would be critical in establishing a spatial reference frame. What are its components? What distinguishes one frame from another? Are there infinite numbers of frames or are there only a few? To address these questions, I will begin by appealing to analytic geometry. The x and y axes in Figure 2.1 are part of a very familiar structure and represent a space in which every point can be defined in x, y coordinates in a two-dimensional (2-D) space. A 3-D coordinate would add a z-axis and a third dimension, but for simplicity the 2-D coordinate will be used here. By frame of reference, I simply mean what others have already specified, namely, a set of reference standards that on a set of coordinates define an origin (where the axes intersect), axis orientation, and sense of direction, or a positive and negative value (see Palmer, 1999). Evidence for the neuropsychological importance of each of these factors will be explored in the sections that follow, but first it will be useful to examine how frames of reference have influenced the study of visual perceptual organization.

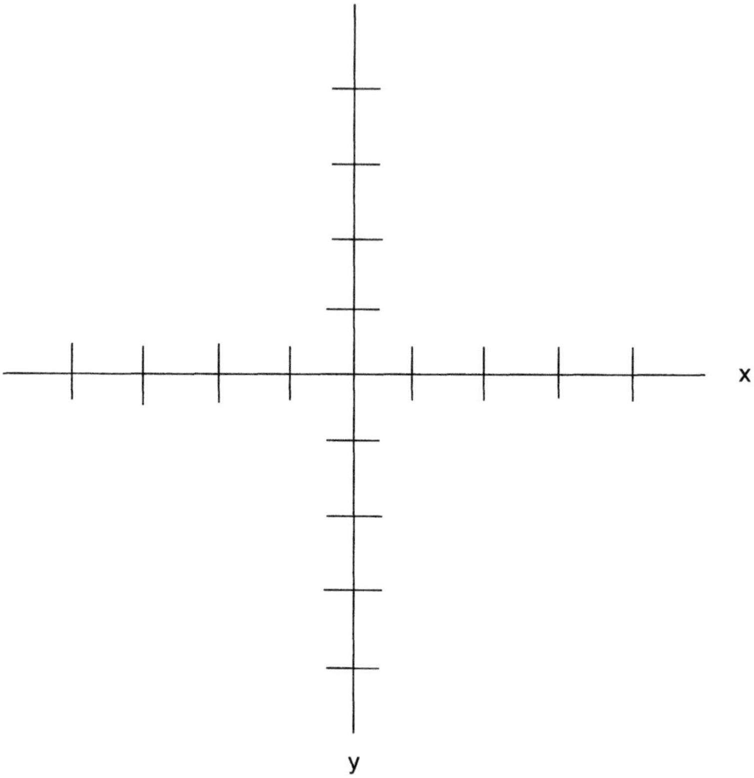

FIGURE 2.1. Typical 2-D Euclidean spatial coordinate. The origin is at the center where the axes cross, and up and right are positive. The smaller marks represent unit size.

A Hierarchy of Reference Frames

The introduction of spatial frames of reference to account for certain perceptual phenomenon was made by the Gestalt psychologists in the early part of the last century (Koffka, 1935). In their tradition of using phenomenological methods, they supported their hypotheses by simply providing visual examples, so that everyone could see for themselves what perception could do. For instance, the example on the right side of Figure 2.2 (Kopferman, 1930) was used to demonstrate that the perception of a shape (on the left) could be changed by enclosing it in a greater whole. When viewed alone, the pattern on the left is perceived as a diamond, but when viewed within the tilted rectangle, the same shape is perceived as a square. The frame of reference that defines the global form changes the perception of the local part by changing the spatial orientation of the local part relative to the whole.

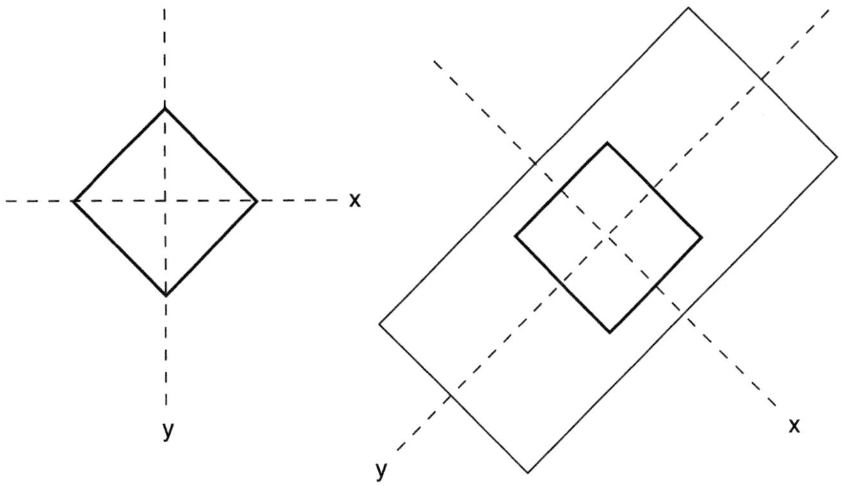

FIGURE 2.2. Kopferman figure showing a single shape that is typically perceived as a diamond (left) with perception of the same shape becoming a square (right) when a rectangle slanted 45° surrounds it, transforming the spatial coordinates in accordance with elongation of the rectangle.

The role of frames of reference in recognizing shapes was later explored more objectively and in greater detail by Rock (1973). In several experiments he showed that shapes presented in one orientation were not as likely to be remembered when they were later presented in another orientation. Similarly, the shape in Figure 2.3 may not be recognized as a geopolitical entity until the page is turned 180°. The default reference orientation is upright and aligned with the viewer or the page, and the shape in the figure is not recognized until the reference coordinates are rotated 180°.

The clear need for some sort of spatial frame of reference in shape recognition has also had enormous influence on computational accounts of object perception (Marr, 1982) and perceived shape equivalency (Palmer, 1999). Such frames provide the spatial skeleton for the creation of computational systems that mimic human perception. Perhaps due to the long history of interest concerning the role of reference frames in object perception, these are often referred to as "object-centered frames of reference," rather than spatial reference frames. Their name likely derives from the fact that the influence of frames of reference has been studied mostly within investigations addressing how we perceive simple shapes as objects or simple clusters of shapes as grouped within a unified frame of reference (Palmer, 1980).

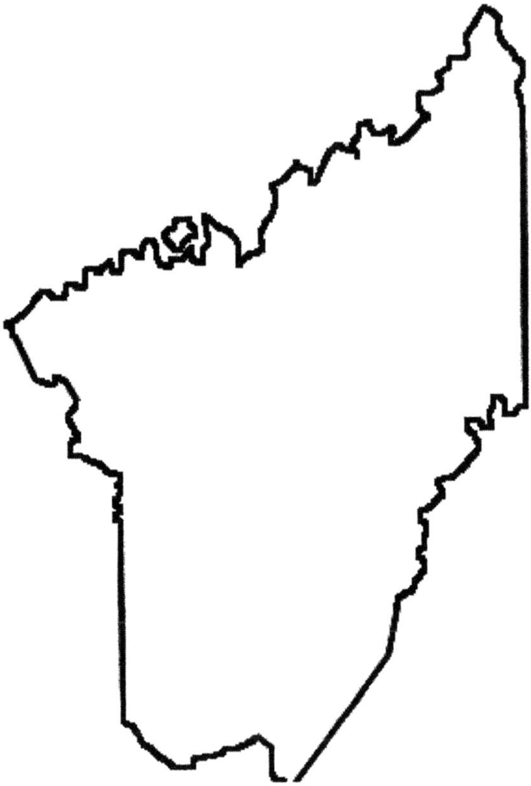

FIGURE 2.3. What state in the United States is this? If you do not know, turn the page upside-down.

Interactions Between Frames

Another area where the effects of frames of reference have received a great deal of study is that of perceptual illusions, as in the well-known rod-and-frame effects that were initially investigated by Asch and Witkin (1948). When presented with a simple bar, adjusting the bar to vertical was influenced by the orientation of a rectangular shape placed around it (Figure 2.4). Asch and Witkin asked their subjects to orient the bar within the rectangle to gravitational vertical while sitting in a completely dark room. Only the lines of the stimuli were illuminated. When the rectangle was tilted, subjects also tilted the line off vertical in the same direction (clockwise or counterclockwise). This effect has been attributed to object-based frames provided by the surrounding rectangle. The larger object (in this case the rectangle) defined a frame of reference within which the line

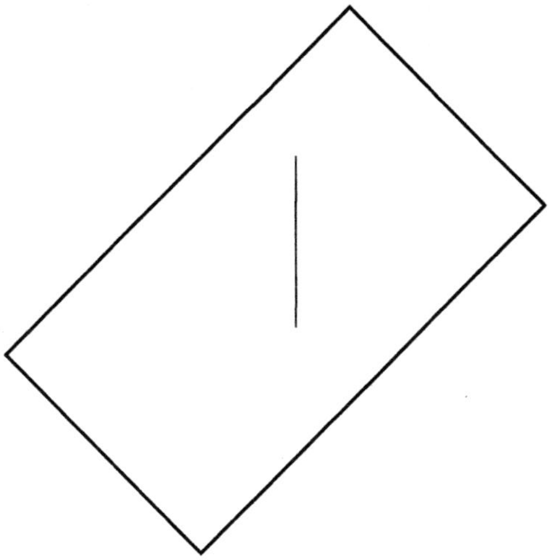

FIGURE 2.4. Example of a rod and frame stimulus in which a person might be asked to adjust the center bar (line) to upright. (Adapted from Asch & Witkin, 1948.)

was processed. Spatial coordinates centered on the rectangle in Figure 2.4 would define an origin where x and y axes intersect (the center of the rectangle), an orientation that is 45° from viewer upright (which becomes 0° upright in the tilted object-centered frame), and a reference sense of direction (up as toward the upper right relative to the page and left toward the upper left). When normal perceivers were asked to adjust the line to gravitational upright, the error reflected the larger frame's dominance.

This simple example brings forth many questions. Unlike the perception of the rectangle in the Kopferman figure (Figure 2.2), the bar in Figure 2.4 is not completely dominated by the rectangle, but the rectangle does influence the bar's perceived tilt somewhat. If only the selected frame of the rectangle defined coordinate referents in Figure 2.4, why is the line not rotated to align with the rectangle? Since viewers were sitting upright in a chair looking at a display in a dark room, the pull of vertical must have come from either the viewers themselves or gravity. In fact, both seem to play a role in performance on the rod-and-frame task and to interact with the global frame of reference. In a more recent study Prinzmetal and Beck (2001) manipulated the orientation of the rectangle orthogonally with the orientation of the viewer (using a tilting chair) and found influences of both viewer-centered and gravity-centered referents as well as an influence of the global frame itself (i.e., all frames interacted).

Viewer-centered, or what are sometimes called egocentric, reference frames are those in which the viewer's body defines the spatial referents. Within viewer-centered coordinates, the reference *origin* is most often fixation but could also be any point along the vertical axis of the head or torso. The reference *orientation* is the axis running through the body midline from feet to head, and the *sense* (of direction) is defined by the head as up and feet as down and right and left relative to the direction the viewer is facing. Gravitation-centered reference frames are those defined by gravity with the sky above and the ground below. The vertical axis intersection with a point along the earth's horizon may act as the reference origin. So, in addition to the multiple frames that capture the hierarchical structure of the visual world, there are additional frames that describe invariant spatial properties provided by gravity and the body itself. All of these frames may be structured into subordinate spatial hierarchies. As Figure 2.5 demonstrates, there is not just one spatial frame centered on the body. An arm has its own spatial coordinates, as does a leg or foot, but each local frame is spatially related to each other within the more global reference frame.

A hierarchy of different gravitational frames that encompasses the universe could no doubt be configured as well (especially by physicists or astronomers who spend their time contemplating the structure of outer

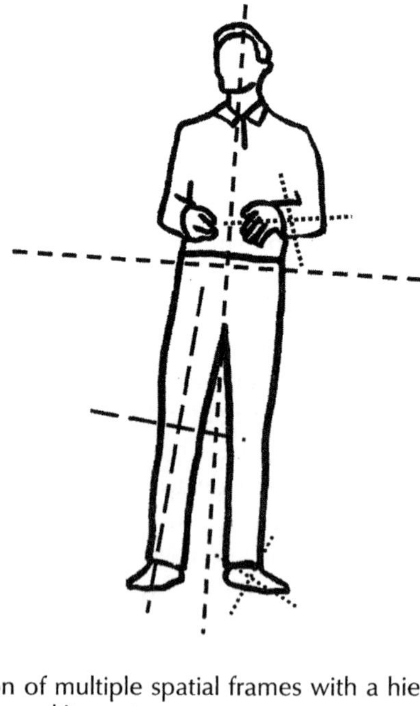

FIGURE 2.5. Cartoon of multiple spatial frames with a hierarchical spatial structure centered on the body and its parts.

space), but most perceptual experiences are centered on the earth, so I will dispense with gravitational frames beyond earth–sky boundaries.

Last but not least, there is the frame of the eye itself (retinotopic space), which tends to dominate vision research in the neurosciences. However, much more will be said about the spatial coordinates defined with reference to the eye and their correspondence to cortical maps when such maps are discussed in a later chapter. For now, I will limit my comments to what I will loosely classify as object-based, viewer-based, and environment-based or scene-based, frames of reference (of which gravity is a special case).

Object-centered Reference Frames

Palmer (1999) defined an object-centered reference frame as "a perceptual reference frame that is chosen on the basis of the intrinsic properties of the to-be-described object" (p. 370). But what are the "intrinsic properties" of an object that influence reference frame components? As Palmer himself pointed out, if we cannot articulate these properties, then the definition is not very useful. Fortunately, Palmer and others have spent a great deal of time investigating what these properties might be.

When establishing the referent orientation of any object, elongation, symmetry, and a base or bottom that defines the ground seem to be important (Figure 2.6). Consider an equilateral triangle (Palmer, 1980) such as that in Figure 2.7a. In the perceptual world, it does not point in three directions at once. We see it point either right, downward toward the left, or upward toward the left. Its direction may appear to change abruptly, but we don't see it point in all three directions at the same time. In fact, normal perceivers have a bias and more often see the triangle pointing right than pointing in one of the other two directions (Palmer, 1980).

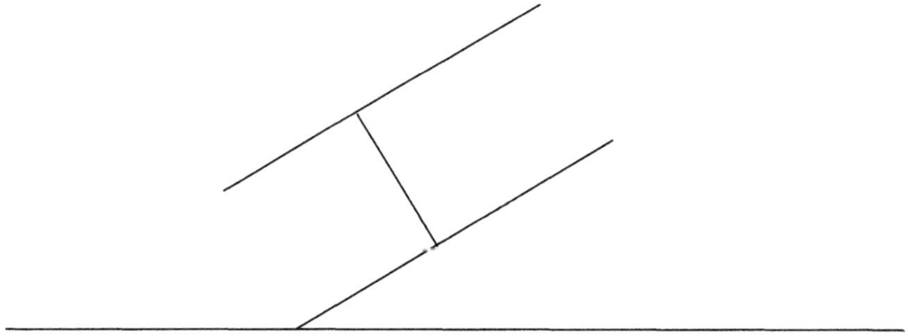

FIGURE 2.6. The H appears unstable and ready to fall within the frame of reference defined by the horizontal line interpreted as ground.

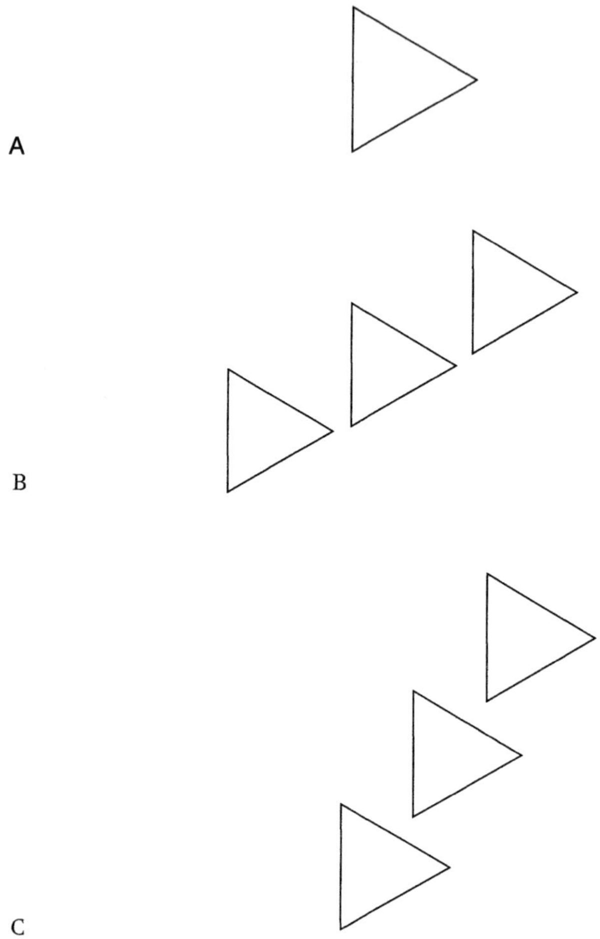

FIGURE 2.7. An equilateral triangle (a) is perceived to point in one of three orientations, but not three orientations at the same time. Placing two triangles around it to form a base biases perceived pointing in the direction perpendicular to the base (b), while placing two triangles aligned with an axis of symmetry biases perceived pointing through that axis (c). (Adapted from Palmer, 1980.)

When other items are added, such as two triangles aligned to produce a base as in Figure 2.7b, all the triangles are then more likely to be seen as pointing perpendicular to the base (upward and to the left in the figure), but when the three triangles are aligned as in Figure 2.7c, they all are more likely to be seen as pointing through an axis defined by elongation and global symmetry (downward and to the left). As long as there are no properties that conflict with other potential frames of reference, reference orientations provided by the environment or the viewer will "win" by default, but elongation, symmetry and base stability can change the referent orientation, as it does in Figure 2.7.

Using a rather different method, Rock (1983) demonstrated that environmental axes were dominant when certain intrinsic properties that define a reference frame were not present in the stimulus (see Palmer, 1999). Rock (1983) presented a shape like one of those in Figure 2.8 and later asked participants to recognize whether they had seen it before when presented in a different (left and middle pictures in Figure 2.8) or in the same orientation as first shown (left and right pictures in Figure 2.8). Recognition was better when the shapes were presented in the same orientation in which they were first seen. This occurred even when viewers were tilted so that the retinal orientation corresponded with the pattern as it was first presented (tilt your head right to see the effect). The environment rather than the viewer was the default frame of reference when competing intrinsic object-based properties were not available (e.g., Figure 2.7).

Another study (Wiser, 1981) showed the importance of elongation and base stability by performing a similar experiment with shapes like those in Figure 2.9. This time the elongated shape with the base tilted was presented first (the right picture in Figure 2.9), and later it was presented again either tilted or upright on the page. Now, people were just as good at recognizing the shape as the same as the one they first saw when it was in upright orientation as when it was in the orientation as originally presented. Rock (1983) argued that the perceptual system stores such shapes in a gravitational framework by defining the base as ground thus overpowering intrinsic object coordinates.

Figure 2.10a shows the originally presented shape overlaid by coordinates that place the x axis at the base and the y axis through the symme-

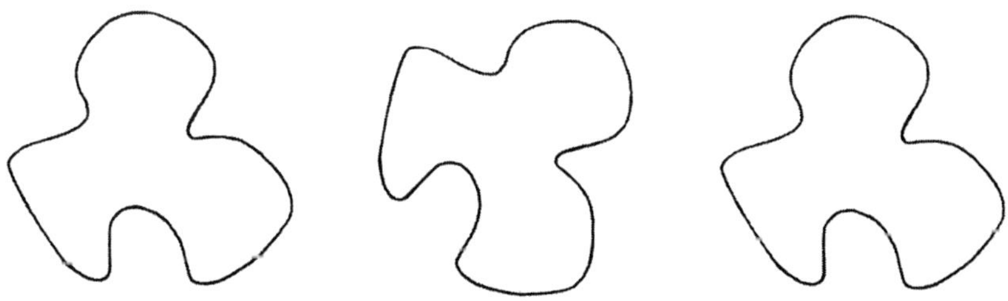

FIGURE 2.8. If the shape on the left is presented and a normal perceiver is later asked to determine whether the shape in the middle or the shape on the right was presented, the perceiver will be more likely to choose the one on the right with the same orientation even when their heads are tilted clockwise 45° to align with the shape in the middle.

FIGURE 2.9. If the shape on the right is presented and normal perceivers are later shown the shape on the left, they are as likely to recognize it as when the shape is shown in the same orientation as initial presentation.

try of the figure; the orientation of the object is positive from the origin (defining upward as perpendicular to the base). Object-based reference frames in this examples is coded "as if" the object were upright in gravitational coordinates. Notice that if only intrinsic properties of the object contributed to shape perception, the x-axis should slide up toward the middle of the shape (Figure 2.10b), changing the origin and also changing the sense of direction for the bottom half. The base of the object would then be downward rather than defining the horizon or ground that could hold the shape stable.

But there is still something missing in these examples of shape-based effects. If there are frames that define spatial properties of objects and frames that define spatial properties of the environment (or what Rock often referred to as gravity-based frames because they followed the laws of gravity), where in the hierarchy does an object-centered frame become an environment- or gravity-based frame? Is the page surrounding a stimulus the environment or is it another object? This question has never been satisfactorily answered to my knowledge.

For this reason I will adopt a rather different view of reference frames, where each level of the perceptual hierarchy is defined by a spatial coordinate in which individual units (e.g., parts, objects, groups, etc.) may or may not be objects but are organized into spatially related units by a hierarchy of frames (from the cushion of my chair to the view off my deck). In this way the frame of reference that defines the spatial referents for the words on this page has the same conceptual status as the frame that de-

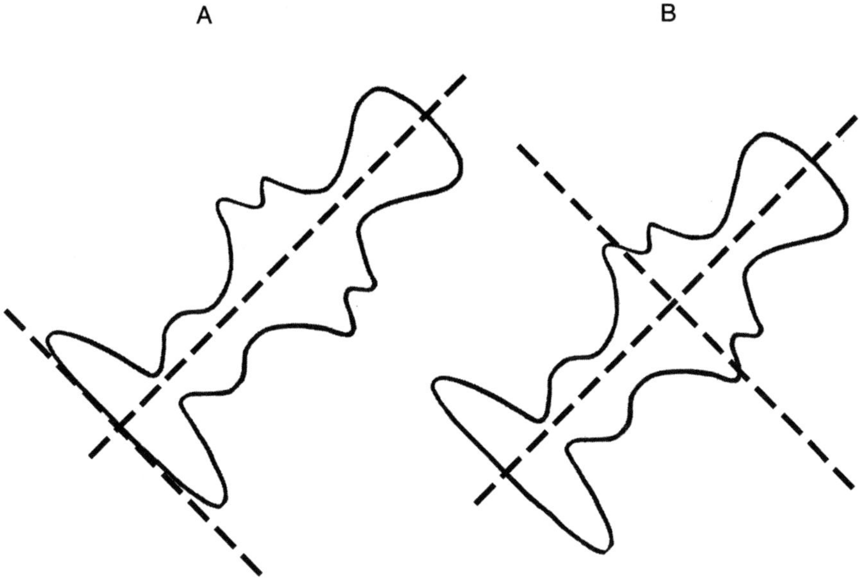

FIGURE 2.10. If the origin of the reference frame intrinsic to the object were placed at the base, this would suggest a base sitting on a ground that is stable (a), while an origin that was centered at the center of the object (b) would defy this principle.

fines this word. Each has an origin, a referent orientation, spatial scale, and sense of direction (see Logan, 1996, for a similar view).

But is there any evidence that our brains respond to these aspects of reference frames that are anywhere like spatial coordinates of analytical geometry? To address this question, the critical components, namely orientation, origin, and sense of direction will be discussed separately in the following three sections. The component of unit size is more problematic and will be discussed later in the chapter.

□ Origin

For a coordinate system to be invoked, there must be a point of origin (or an intersection where axes cross). In retinal coordinates the origin is the point of visual fixation, and this point is also where attention is typically focused. When my eyes are looking forward, but I am attending to something in my peripheral vision, fixation and the locus of attention are dissociated. One question becomes whether the attentional locus can act as an origin for a frame of reference, and the answer seems to be yes. This has ramifications not only for studies of normal perception, but also for how to interpret many studies of attention in cognitive neurosciences.

A Case Study and Reference Frame Origin

Some intriguing evidence concerning the structure of spatial frames comes from a case studied by Michael McCloskey, Brenda Rapp and their colleagues (McCloskey et al., 1995; McCloskey & Rapp, 2000). They tested a person (AH) with abnormal spatial abilities who often perceived a briefly presented stimulus to be in the mirror image location about a vertical or horizontal axis (i.e., reflectional symmetry). AH was a successful college student at a major university when she was tested, and not a patient in any sense. But she perceived some spatial peculiarities that have a great deal to say about the structure of spatial reference frames and the role of attention in determining the origin of frames. For this reason I will discuss her performance in some detail.

Several studies of spatial location abilities were reported with AH, but the most relevant ones for the present purposes have to do with the type of location errors she made. When AH was presented with an item at one of four locations horizontally aligned across a computer screen (Figure 2.11a), her location errors were systematic (I will label the stimulus locations at which a target could appear as P1, P2, P3, and P4). P1 and P4 were mirror image locations around the vertical axis through fixation, as were P2 and P3. In another condition, stimulus locations were aligned vertically with mirror image locations then defined around the horizontal axis (Figure 2.11b). The common origin in these two cases was fixation. The same pattern of performance occurred in both conditions and was reflectionally symmetric. Location errors for stimuli presented at P1 were consistently misperceived as in the position of P4, and location errors for stimuli presented at P2 were misperceived as in the position of P3

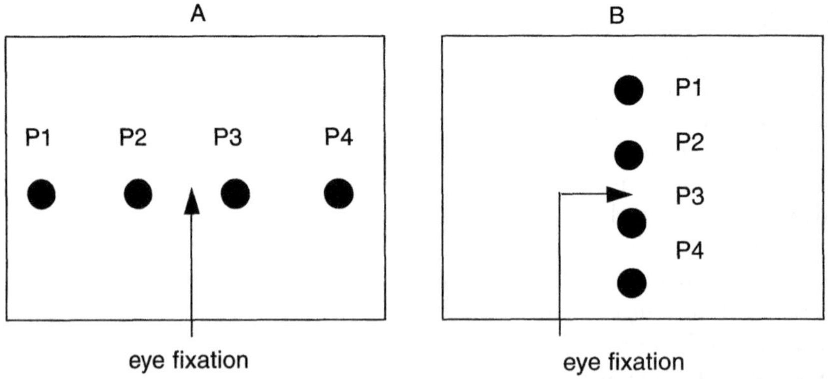

FIGURE 2.11. Position (P1, P2, P3, P4) placed symmetrically around fixation horizontally (a) and vertically (b).

(McCloskey & Rapp, 2000) and vice versa. Where AH saw the stimulus and where it actually was located were in symmetrically opposite locations. Her errors were not random, as would be expected if she simply forgot the stimulus location or had double vision or if she had no structural space at all. Rather, errors in localization could be predicted from the location of the stimulus itself in this case relative to fixation..

This study did not establish whether these effects reflected polar coordinates, as might be expected in retinal space or Cartesian coordinates, which might be more influential in environmental space, nor did it address the question of attentional fixation as an origin. However, an earlier study that required AH to ballistically reach for objects on a table in front of her showed that her location errors were not represented by polar coordinates (McCloskey et al., 1995). First causal observation suggested that all of her location errors were left/right or up/down but not diagonal. This prompted a study in which 10 stimulus locations forming two arcs were sampled (Figure 2.12). On each trial a small cylinder was placed at one of the locations represented by the dots. Half of the locations formed an arc 18 cm away (close) from AH and half formed an arc 36 cm away

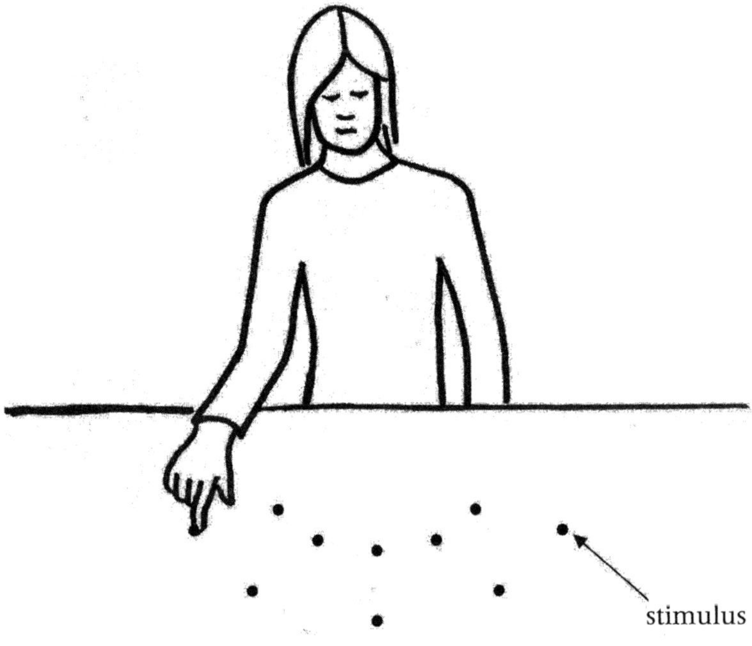

FIGURE 2.12. Position of a participant (AH) reported by McCloskey, Rapp, and colleagues and the locations on a table where a stimulus could appear (represented by the dots on the table top). Her ballistic reaching direction was in the symmetrically opposite location from where the stimulus appeared.

(far). The critical conditions were the 8 locations to the left and right of her vertical midline. For these locations AH made location errors about two thirds of the time, and in every case her errors were mirror image errors. For close locations her errors were always close and in the mirror image location. Likewise, for far locations her errors were always far and in the mirror image locations. These findings show that her distance perception was accurate (the spatial scale factor was intact). Even though she would reach toward the wrong side, her movements were to a correct distance from her body. What was most impressive was that all of her errors showed reflectional symmetry around an imaginary vertical axis through the middle of the display, which was aligned with her body midline. AH did not reach for a diagonal position from the cylinder's location as would be expected if she represented space in polar coordinates. Rather, all her reaching errors could be described by a Cartesian frame of reference.

Is the Origin the Origin of Attention?

None of the studies I've discussed so far have dissociated AH's fixation or body midline from the location where attention might be located. Does attention play a role in establishing the origin of a spatial frame or is it the relationship between body midline, fixation and environmental coordinates that define spatial locations in the field? To address this question, McCloskey and Rapp (2000) dissociated eye fixation location and attentional location, again using the stimulus locations as shown in Figures 2.11a and 2.11b. They first directed attention to an intermediate location between P1 and P2 or between P3 and P4 to measure whether AH's location errors would be predicted by an axis defined by the center of attention or by an axis defined by fixation. In order to assure that AH kept her eyes fixated at the center of the display, eye movements were monitored, and in order to encourage her to attend to the intermediate locations (between P1 and P2 or P3 and P4), a variable number of small dots were briefly presented there and she was instructed to report the number of dots on each trial.

The question was whether her location errors would be the same as they were in the previous experiments (supporting an account in terms of a body-centered frame or eye fixation) or be symmetric about the focus of attention. The results were very clear. All errors were around the focus of attention. For example, when an item was presented at P2, her errors were to P1 and none were to P3 or P4, and when an item was presented at P3, her errors were to P4 and none to P2. This pattern was evident both when the locations of the presented items were vertical and when they were horizontal. The locus of attention determined the origin of the spatial frame of reference.

Origin and Center of Mass

The previous studies demonstrated that volitionally directing attention to a location influenced the reference frame over which location errors occurred. But attention typically does not linger at a given location for any great length of time. Generally attention moves through the world seeking the most critical information for the task at hand or is pulled to some item or location automatically by, for instance, detecting salient changes such as an abrupt onset, movement, or novel event (see Yantis, 1993). A bolt of lightening that occurs anywhere within the visual field is likely to attract attention to its location. A sudden movement along the wall might attract attention, and an eye movement may rapidly follow to determine whether or not it is a spider. The movement's location is detected, but an eye movement is needed to determine what the object might be.

Even in static displays there are properties that will attract attention to a location, and at least one, the *center of mass* is also influential in determining where fixation will land after initiation of a rapid eye movement or saccade. Saccades to a salient target overshoot when irrelevant items are presented beyond the target in the periphery (Figure 2.13b) and undershoot when items are presented between the target and current fixation (Figure 2.13a) (Coren & Hoenig, 1972). The center of mass of the

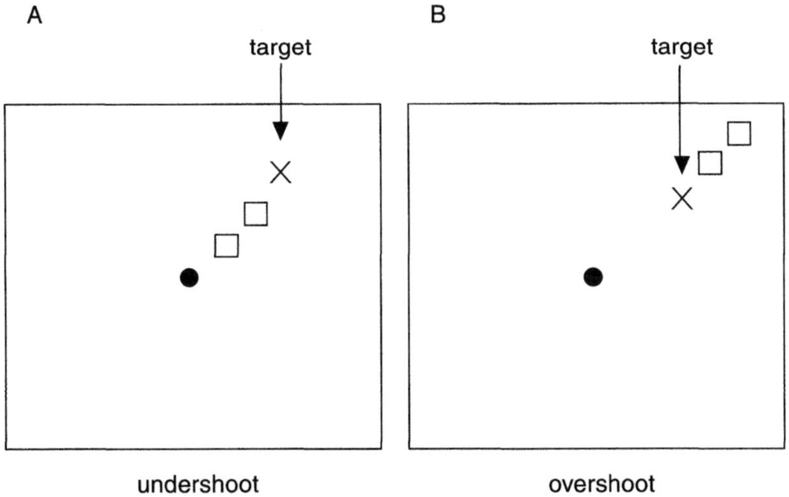

FIGURE 2.13. When instructed to make a rapid eye movement to a target (X) in display a, eye movements tend to undershoot, but when moving to a target in display b, they tend to overshoot. The center of mass in a and b influence eye movements.

stimulus array pulls the target location for a saccade in one direction or another. Overshooting or undershooting can be overcome by volitional control, and this is interesting in its own right, but the most important question for the present discussion concerns how the origin of a reference frame can be established. As it turns out, attention also responds to the center of mass of a display, indicating early interactions between the distribution of sensory input and establishing the attentional origin.

A postdoctoral student in my laboratory, Marcia Grabowecky, tested this question by exploiting the well-known observation that search time increases as a function of set size when searching for an O among Qs (Figure 2.14). She configured search displays of Os and Qs in arcs that could appear anywhere on a circular wheel around fixation and varied the target position within the arc. For instance, the O could appear in displays like that shown in Figure 2.15, where the O is at the center of mass in one case (a) but not in the other (b). She then measured reaction time for normal perceivers to determine whether an O was present or absent. In all cases, eyes remained fixated in the center of the display

FIGURE 2.14. Example of a search display with a target O and distractor Qs requires a serial search.

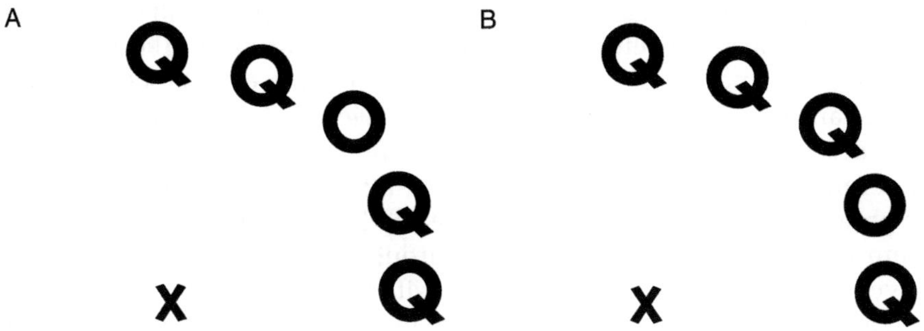

FIGURE 2.15. The target O in the left display (a) is found faster than the target O in the right display (b), presumably because in a the target is in the center of the search display (center of mass), while in b it is not.

where the X is shown in Figure 2.15. The results demonstrated that when the target was at the center of mass, it was detected faster than when it was not.

It appears that attention was drawn to the center of mass of the display where search for the target began. These findings show that where search begins depends on the location of attention as opposed to where the eyes might be at any given moment. The origin of the "object" is the center of the parts of the stimulus defined by their spatial relationships. As mentioned earlier, eye movements are also sensitive to the center of mass, and functional magnetic resonance imaging (fMRI) data have shown extensive overlap between eye movements and attentional movements (Corbetta et al., 1998). These findings have been used to argue for the priority of eye movement planning is directing attention. But the fact that attention is influenced by the center of mass means that the extent of the stimulus display is coded and its center calculated before eye movement direction is programed. Calculations of an origin or center seems to occur first with eye movements following, rather than the reverse. This origin then sets up the frame in which attentional search proceeds.

Center of Mass Attracts Attention in Neglect

The conclusion at the end of the last paragraph was supported in a study by Marcia Grabowecky, Anne Treisman, and myself in 1993, where we addressed how the center of mass might affect visual search in patients with unilateral neglect (also see Pavlovskaya, Glass, Soroker, Blum, & Groswasser, 1997). Recall that these are patients who do not attend to information contralateral to their lesion (see chapter 1). The study included 7 patients with unilateral neglect (for simplicity, I will refer to the neglected side as the left side, which is true in most cases of neglect).

The phenomenon of neglect presents something of a paradox. If neglect can occur in object-based as well as viewer-based coordinates (as was described in chapter 1), how is the center of what is to be neglected determined without a full spatial representation of the display? Suppose I approach the bedside of a patient suffering from neglect and ask the patient how many people are standing around his bed. Suppose first that seven students have accompanied me, with four standing on the patient's left and three on the patient's right with me. In this case the patient might report seeing four people and describe each of us on the right side of his bed. Then suppose that four of us leave (me and the three students standing with me on the right). Now the patient is likely to report seeing two people and describes the two who are on the rightmost of the remaining four (who are still standing on the left side of his bed). How did the visual system establish the center of each group (eight in the first example and four in the second) without first seeing the extent and spacial arrangement of all the people around his bed? Phenomenon like this evoke the existence of some sort of preattentive process that calculates the spatial extent and origin of a visual display before the location of left and right are established. In this way the left items relative to the center or origin are neglected.

In the Grabowecky et al. (1993) experiment, the issue of preattentive registration of spatial extent and the influence of the center of mass were examined in a group of patients with moderate to severe spatial neglect. Although I discussed the example above as if neglect has a clear demarcation down the center of a display, in fact it is far more variable. The distribution of spatial deficits on the contralesional side could be as small as neglecting one or two items on a page (perhaps the ones at the leftmost, bottom, as in Figure 2.16a) or it could be as large as neglecting everything to the left of a few items in the rightmost column (Figure 2.16b). To try to control for this variability, we only tested patients who were fairly similar in terms of the number of items that were neglected on the Albert's line crossing test (Figure 2.16), which is a typical bedside test for neglect. Any patient who crossed out lines on the contralesional side of the page were not included in the study, but all were required to cross out at least the rightmost column so we could be confident they were alert enough to perform the task.

The task in the main study was to find a conjunction target in a diamond-like display. The diamond always appeared in the center of a page and half the time the target was on the right side of the diamond and half on the left (Figure 2.17). We knew from previous research that searching for this type of target on the left (neglected) side was difficult and often

FIGURE 2.16. When presented with a number of lines positioned across a page and asked to cross out every line, patients are diagnosed as having unilateral visual neglect whether they miss only one or two lines (a) or most of the lines (b). The outlines represent missed lines in the two examples.

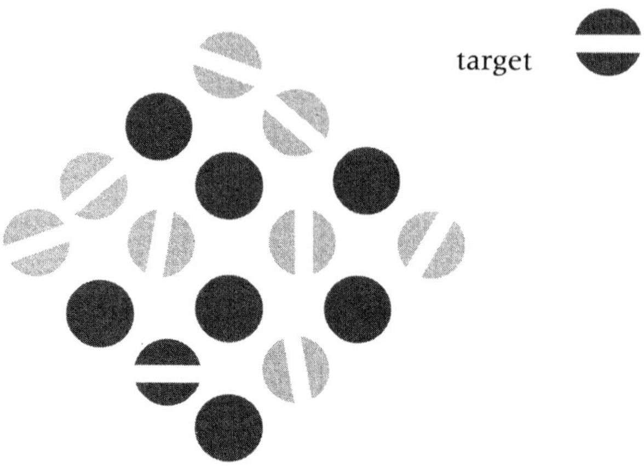

FIGURE 2.17. Example of a display diamond that was used to test visual search in patients with unilateral neglect. The groups of circles were centered in the middle of the display and patients' response times to find the target were recorded. The target example was not shown with the display diamond. (Adapted from Grabowecky et al, 1993.)

took several seconds (Eglin et al., 1989). We also knew that patients would continue to search as long as they were confident that a target was present in every display (perhaps cuing themselves in some way to move leftward when the target was not found on the right side—a common rehabilitation technique with these types of patients). We first replicated the "contralateral delay" that Eglin et al. (1989) reported. It took a bit over four seconds on average to find the target when it appeared on the left side of the diamond, but only about 1.5 seconds when it appeared on the right side.

The center of mass was then manipulated by adding irrelevant flankers to the left, right, or both sides of the centrally positioned diamond, and response time to find the target was again recorded. When flankers were present on only the right side of the diamond (Figure 2.18a), search time to find a target on the neglected (left) side increased to about 12 seconds (i.e., left neglect became worse), as shown in Figure 2.19. But the most impressive finding was that when flankers were added to both sides of the diamond (Figure 2.18b), detection time returned to around 4 seconds. These findings show that the time to find the target in the diamond was not due to the number of items on the right side that could attract attention, but to something else that took into consideration the balance between the two sides. We suggest this "something else" is the center of mass that modulates the rightward bias by changing the origin of attention. A comparison of search time to find the target under conditions

A

FIGURE 2.18. Example of irrelevant flankers placed on the right side of the display diamond (a). The mirror image of this figure was also presented in which flankers appeared on the left side. Note that the search diamond is the same as that in Figure 2.17. When flankers were placed on both the right and left sides the stimuli looked like that shown in (b). (Adapted from Grabowecky et al., 1993.)

when the center of mass was the same (i.e., when no flankers were present—Figure 2.17) to that when flankers were present on both sides (Figure 2.18b), demonstrated that the degree of neglect was nearly the same when the origin was the same.[2]

These findings are consistent with the observations in normal perceivers showing that the center of mass of a display can pull attention in one direction or another. The center of attention (the origin of a reference frame that defines left and right) is changed by the center of mass as opposed to the amount of information on one side or the other. In this way, the left side of a spatial frame with an origin defined as the center of attention rather than eye fixation is neglected. If the origin defined a single object or perhaps a group of items, this would likely be categorized as object-centered neglect, but perhaps a more parsimonious way to de-

B

FIGURE 2.18b.

scribe this phenomenon is in terms of neglect within a selected spatial frame with a baseline origin that is shifted. Since neglect is more likely in patients with right than left hemisphere damage, the shift is generally in the rightward direction.

In sum, for normal perceivers, search begins at the center of mass and then is equally likely to move to one side or the other. But for patients with neglect, the center of attention, and thus the origin of a reference frame, is shifted ipsilesionally (i.e., rightward in left neglect, and leftward in right neglect) as was seen even when no flankers were present in the displays used by Grabowecky et al. (1993). Irrelevant flankers shifted this baseline bias even further into the right field, but attention was brought back to baseline when bilateral flankers were added and defined the center of mass the same as when no flankers were present.[3]

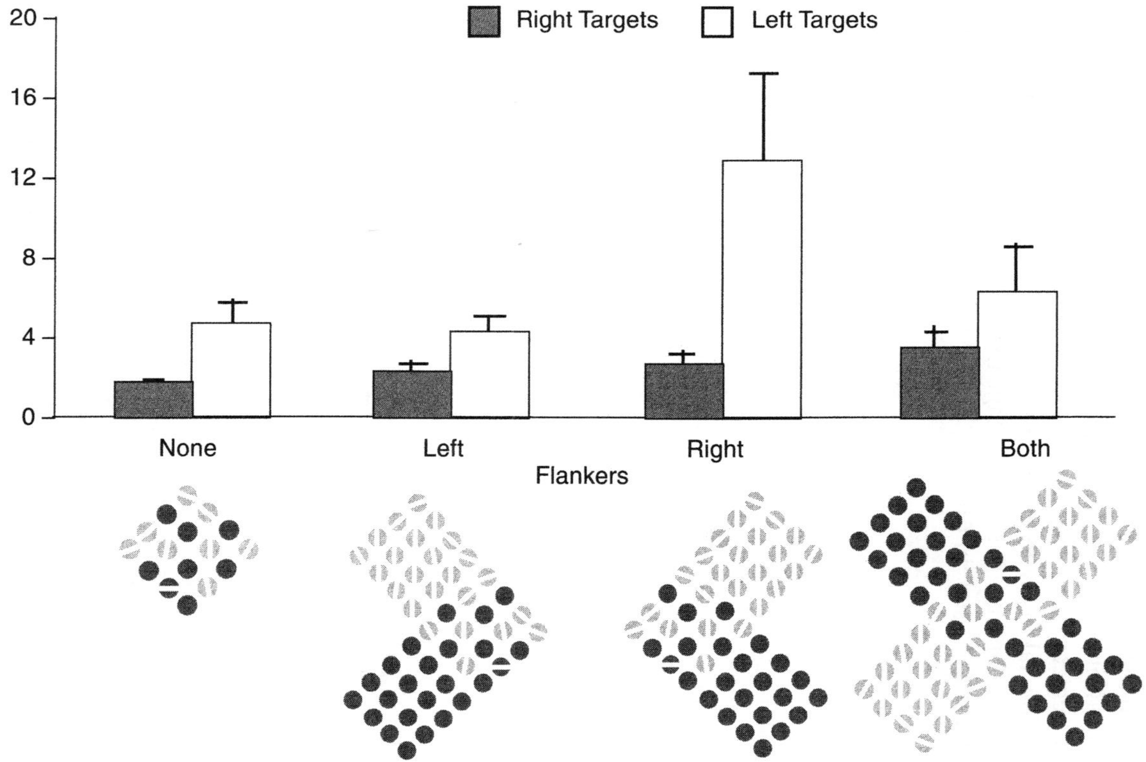

FIGURE 2.19. Mean reaction time as a function of flanker condition when no flankers were present (None), when flankers appeared on one side (Left, Right) and when flankers appeared on both sides (Both). Gray bars are for right-sided targets and white bars are for left-sided targets within the display diamond. (Adapted from Grabowecky et al., 1993.)

These findings are consistent with other findings suggesting that the origin of the reference frame that defines displays as a whole (what are typically called object-based) is placed at the locus of attention. In patients with neglect, this locus appears to be abnormally shifted to the ipsilesional side, taking with it the origin of what is left of the frame after unilateral brain damage. Areas of damage that are most likely to produce neglect will be discussed in Chapter 7.

Orientation

Orientation is another basic component necessary for reference frame representation, and it is massively represented in the visual system. Cells exhibiting orientation tuning first appear in primary visual cortex (DeValois & DeValois, 1988; Hubel & Wiesel, 1959) with a large number of cells in areas further along the visual pathways continuing to prefer certain orientations over others. For instance, motion cells in area MT (see Figure 2.20) respond when movement is in a particular direction, and color cells in area V4 respond more vigorously to a perferred color when it is on a

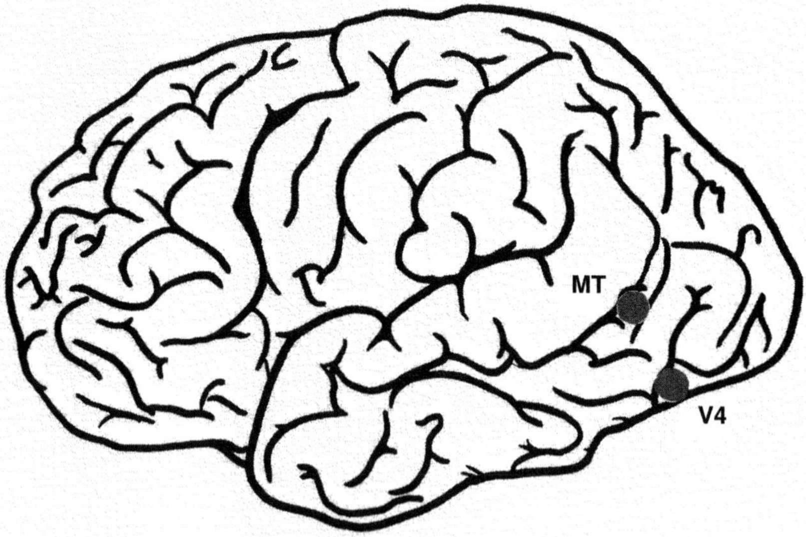

FIGURE 2.20. Example of two of many areas of the cortex where preferences for different features of visual stimuli have been observed. MT is sensitive to motion, while V4 is sensitive to color. V4 is usually on the underside of the brain proximal to the area noted.

bar of one orientation or another (Desimone & Shein, 1987; Desimone & Ungerleider, 1986).

Cells that are orientation-selective fire more frequently to a preferred stimulus orientation with a gradual falling off as orientations deviate from the orientation the cell prefers. In other words, there is orientation tuning. Some cells are narrowly tuned, responding to only a small range of orientations (Figure 2.21a), while others are widely tuned, responding to a large range of orientations (Figure 2.21b). Given the billions of neurons that show orientation tuning, it is clear that the physiology of the visual cortex contains the necessary architecture to rapidly and precisely determine the orientation of stimuli in the visual field and at various levels of spatial resolution.

At the level of gross anatomy, the neuropsychological evidence from patients with brain lesions demonstrates a double dissociation between orientation and location deficits with damage to slightly different areas of occipital-parietal cortex (see De Renzi, 1982; McCarthy & Warrington, 1990). The codes to spatially locate an item and to determine its orientation appear supported by separate neural mechanisms. Patients can lose the ability to visually perceive an object's orientation but continue to locate it (see Milner & Goodale, 1995), while other patients can lose the perception of large areas of space (e.g., extinction or neglect) without losing orientation information of the items they do see. In other words, the various components of spatial reference frames can break down independently, producing representations of stimuli that are perceived normally except for their orientation on the one hand and their location on the other.

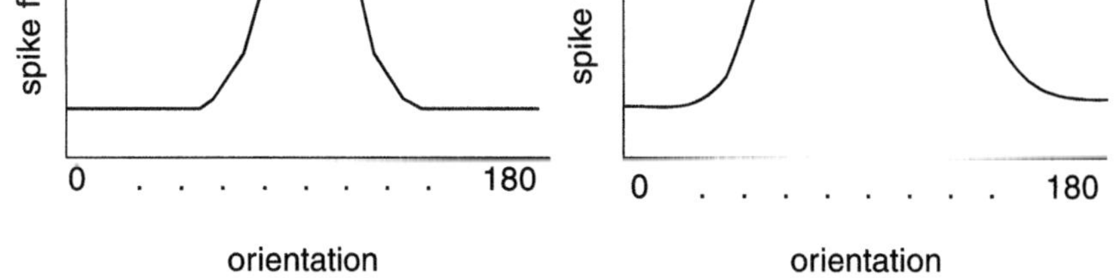

FIGURE 2.21. A cell that responds to different stimulus orientations as in (a) is said to be narrowly tuned, while one that responds as in (b) is more broadly tuned.

Losing Orientation Without Losing Location

One intriguing report of a patient with bilateral lesions in both ventral occipital-temporal lobes was described by Goodale, Milner, Jakobson, and Carey (1991). This patient was unable to correctly match the orientation of lines when viewing them, but her hand movements were guided correctly by orientation. Figure 2.22a shows the orientation errors she made when matching two lines visually, while Figure 2.22b shows the errors she made in hand orientation when asked to mail a letter through a slot that varied in orientation. Both figures are standardized to vertical and plotted according to the angular errors she made (i.e., the difference between the stimulus orientation and her response). The differences between visual matching and motor matching are striking. Visual matching was completely disrupted, while motor matching was intact. In the language of reference frames, one could describe the results as a deficit in matching orientation within extrapersonal frames in the visual task with intact reference frame alignment in a viewer-centered frame of reference in the motor task. Also revealing is that both visual and motor abilities to *locate* lines remained intact.

The dissociation between visual perception and motor control is interesting in its own right and its implications and influence will be discussed more fully later. However, the important point here is that the perception of *orientation* in object- or environment-based frames was severely affected

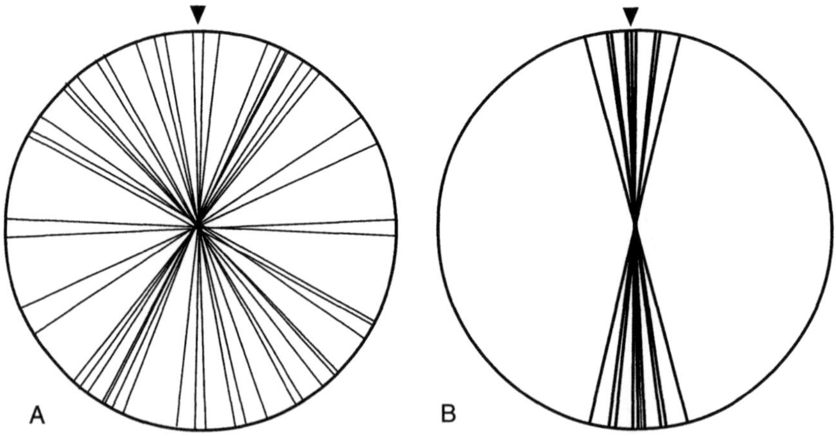

FIGURE 2.22. Orientation disparity between orientations presented (represented by vertical) and what was reported by a patient with bilateral ventral occipital-temporal lesions (a). The same patient's performance when an envelope was placed in her hand and she was asked to drop it through a mail slot (b). (Again, the orientation of the mail slot is normalized to vertical). (Adapted from Goodale et al., 1991.)

by lesions that disrupted ventral stream analysis, while viewer-based frames appeared to have remained intact. In addition, it was only the frame component of orientation that showed this dissociation. Localization was not affected, something that has been mostly overlooked in discussions of the Goodale et al. (1991) findings.

These behavioral findings are reminiscent of those reported by Warrington and Taylor (1978) in a patient who could not identify objects in noncanonical orientations. When a familiar object was placed in an orientation in which it was most often seen, identification was rapid. But when it was rotated into less typical orientations, identification failed. This patient, too, had occipital-temporal damage.

Other investigations have focused more on orientation and occipital-parietal function, especially in the right hemisphere. A very common test used to assess orientation perception is one in which a standard is given either simultaneously or sequentially with a set of orientations and the patient is asked to select the orientation that matches the standard (Benton, Varney, & Hamsher, 1978). For instance, in Figure 2.23 the line at the top (standard) is the same orientation as number 7 in the radial below. Some patients with right occipital-parietal damage find this task especially difficult but may be able to match the location of dots on a page quite well. A

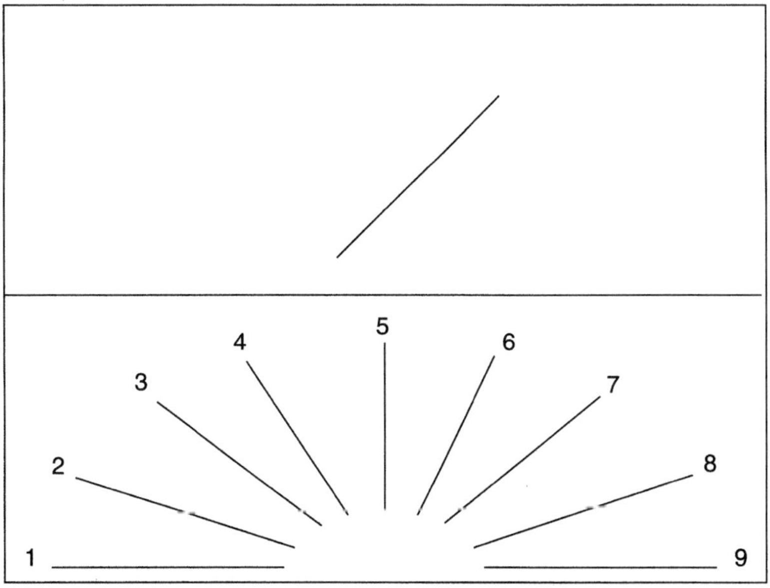

FIGURE 2.23. A neuropsychological test in which patients are shown a single line that could be oriented like that on the top and asked to choose the line at the bottom that is in the same orientation.

3-D version of the orientation matching test was developed by De Renzi, Faglioni, and Scotti, (1971), and basically showed a similar distribution of the lesions that disrupted orientation matching with 2-D stimuli. Although damage in this area may also affect functioning in ventral areas as well, it is clear that lesions restricted to parietal areas can disrupt orientation perception.

Orientation and Normal Perception

Orientation representation has been extremely influential in several theories of perception and in computational accounts that have had reasonable success at modeling object vision. For instance, Marr (1982; Marr & Poggio, 1979; Marr & Ullman, 1981) developed a detailed and influential computational model of how a 3-D percept of an object could result from a few fundamental descriptors, with orientation being a critical component. The central role of orientation in Marr's model was based initially on the neuropsychological deficits reported by Warrington and Taylor (1978) and discussed above.

Warrington and Taylor's patient suffered from "apperceptive agnosia," or a deficit in the ability to discriminate shapes visually. Although Warrington herself has since suggested an alternative interpretation to that of orientation tuning (Warrington & James, 1986), the basic vertical and horizontal axes necessary in Marr's model of object perception began with the idea that when one axis is foreshortened relative to another (Figure 2.24), the visual system calculates the spatial structure of a stimu-

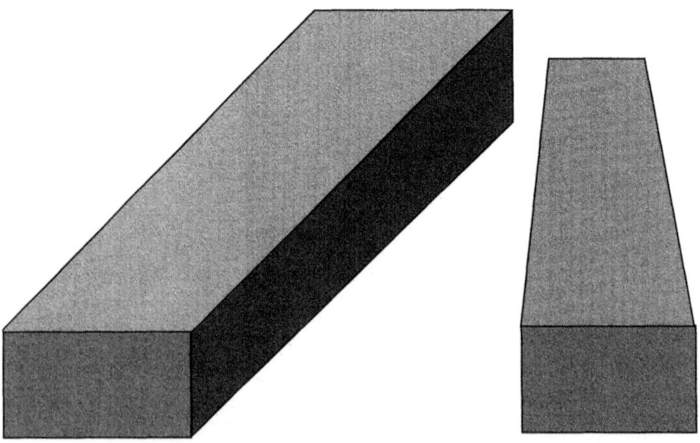

FIGURE 2.24. The figure on the right is a foreshortened version of the figure on the left.

lus within an updated spatial reference frame. If this updating is damaged in some way, then incorrect matches will be made between one object and another rotated in depth. This orientation updating is theoretically independent of mechanisms that determine an object's location, consistent with dissociations described earlier that have been observed in the neuropsychological literature.

Why might orientation be so widely represented in the visual cortex? For one, it provides a critical foundation for the description of perceptual objects (basically a spatial description of primary axes and overall configuration). It also carries information about slant and verticality. In order to see the world as stable and to be able to move around it successfully, the relative orientations of objects and surfaces in a scene must be accurately and rapidly calculated and updated. Parallel processing through distributed systems would be an efficient way to accomplish this basic need.

Again, one can see the necessity of considering a space/object hierarchy with objects linked to each other not only by their relative locations, but also by their relative orientations within selected frames. In Figure 2.25 the orientation of the more global level of the table provides a frame in which the relative orientations of the paper and pencil on the table can be computed. In turn, the paper provides a global frame for the computation of orientation of the words on the paper. The words appear upright in a paper-centered frame, while the paper appears rotated 90° in the table-centered frame.

FIGURE 2.25. The global frame of the table defines the perceived orientations of the items on the tabletop, while the global frame of the paper defines the perceived orientation of the words written on the sheet of paper.

The dominance of the table as a global frame could be attributed solely to orientation selection, but a more efficient way to bind the spatial elements is in some sort of hierarchical spatial organization. There is a great deal of evidence supporting global frame dominance in the literature (e.g., Navon, 1977; Palmer, 1980), and the majority of this evidence suggests that global frames are processed first or more rapidly when all else is equal (e.g., without selective filtering). If the room were to tip, all the objects in Figure 2.25, including the table would tip with it. But if the paper rotated on the tabletop, the room orientation and the table, itself would be unaffected. Nevertheless, the more local elements of the words on the relatively global paper would rotate with the paper. It would not be a violation if the pen rotated in an opposite direction (as it contains its own intrinsic frame separate from the paper), but it would be a violation if the letters rotated with the pen rather than the paper. There are asymmetric links between global and local spatial reference frames that are consistent with the evidence for global dominance. Larger or more global spatial frames provide the spatial structure for the analysis of more local frames, cascading down through multiple levels of object/space.

When considering representations of space and objects, it is therefore useful to think of a hierarchically organized set of reference frames (as Rock first suggested) that operate according to certain principles in spatial coordinate systems. Selection of a location, an orientation, a global or local frame, or any other unit in the visual display is possible, but these may all rely on the spatial reference frames selected at any given moment.

☐ Sense of Direction

To this point, I have discussed evidence for the orientation and origin components of spatial reference frames, but the axes of spatial frames must also be assigned positive and negative values in order to split them into left and right or up and down (i.e., to determine sense or direction). The sky defines up, the ground down. My head defines up, my feet down. The top of an A is at its vertex and its bottom is the plane at the end points of its diagonal lines. This is true whether the A is tilted on the page or upright. Notice that objectively speaking, up and down could be reversed. That is, my feet could define up and my head down in spatial coordinates, but the important point is that the sense of direction represents opposed values along axes that cross through the origin. This might be best exemplified with left and right. Most right-handers label right as positive and left as negative while left-handers label in the opposite way. In either case, the sense of direction of an axis is positive on one side of the origin and negative on the other.

Reflectional Symmetry

One spatial property that has been used to study sense of direction in perception is reflectional symmetry. Reflectional symmetry simply refers to a set of points that exactly replicate themselves when reflected 180° around a selected axis. An O has reflectional symmetry around all axes through its midpoint. Nevertheless, despite its roundness, we assign one point of the O as up and another as left. The object-based frame of the O contains spatial labels. Reflectional symmetry also occurs when two Os are placed, say, 3° to the right and left of the vertical center of a piece of paper since they align perfectly when the paper is folded in half. What typically is called a space-based frame is in fact the frame centered on the sheet of paper (the more global object). Reflectional symmetry in viewer-based frames would occur if I held both of my hands straight out from my body with my palms facing toward the ground. My left thumb would be in the symmetrically opposite position as my right thumb through an axis centered on the midline of my body. However, if I placed my hands with one hand facing up and the other down, the thumbs would no longer be reflectionally symmetric. Reflection over the vertical axis of my body midline produces a misalignment of the thumbs.

The motor system is exquisitely sensitive to this symmetry (see Franz, 1997). Try circling with one hand and moving up and down with the other. Also recall AH, an otherwise normal person with an altered spatial sense of direction between vision and ballistic hand movements (see Figures 2.11 and 2.12). AH grasped items in the mirror image locations from where they were presented and her mislocation errors were reflected around the center of attention when attention was cued to the right or left of central fixation.

Reflectional symmetry depends on the axis running through the origin of a frame that demarcates the midline. For instance, suppose attention is fixated on the + in Figure 2.26a, the reflectionally symmetric image of the frog is as shown in Figure 2.26b and not its reflection around its own intrinsic axes (Figure 2.26c). If we wanted to create a stimulus in which the frog was symmetric about its own axes, we would have to use a different perspective of the frog (e.g., Figure 2.27). These axes also have a sense of direction in Figure 2.26a and 26b, but symmetry is a special case in which positive and negative have a point-to-point correspondence. This fact affords the opportunity to study where an axis bisects a stimulus as well as its corresponding directional properties.

If the vertical axis of the shape in Figure 2.28 is placed through the center of the diamond, then every point on the right replicates every point on the left (i.e., the positive and negative cancel each other, producing the same form after reflection). However, if the vertical axis is

A

B

C

FIGURE 2.26. Reflectional symmetry of the frog in a is b, not c when the origin of the spatial frame is through fixation (the + sign).

displaced to the right of the shape, the reflectional symmetry is destroyed and the bulk of the figure lies on the left. It is not as if the sense of direction is absent in one and not the other, but reflectional symmetry seems to carry weight in terms of aligning multiple frames of reference within a scene.

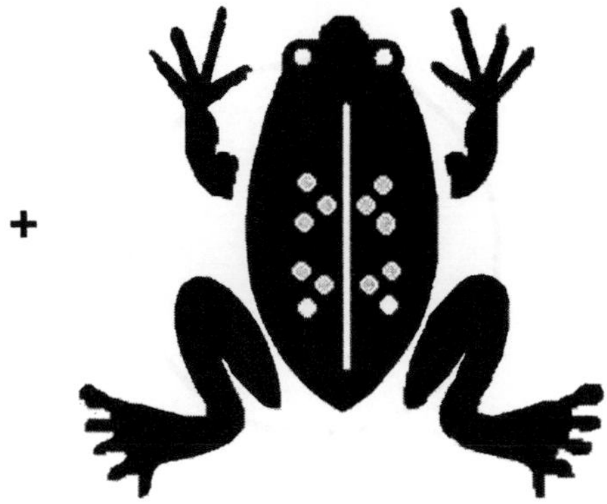

FIGURE 2.27. There is now an axis of reflectional symmetry that is intrinsic to the frog herself.

Reflectional Symmetry and Theories of Object Perception

A great deal of research has been reported in the literature on how reflection influences objects perception. But reflection has generally been defined around an axis with its origin centered on the objects (Figures 2.26a and 2.26c). Some years ago Garner (1974) argued that the "goodness" of

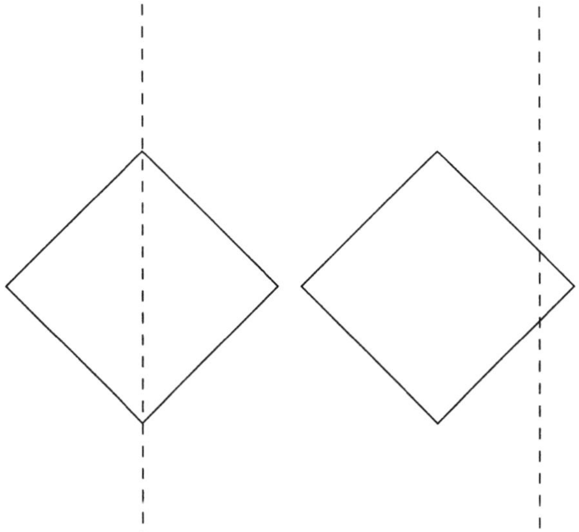

FIGURE 2.28. An axis through the center of the diamond shape, as on the left, produces point-to-point correspondence when reflected, while an axis presented elsewhere, as in the diamond on the right, does not.

FIGURE 2.29. The circle on the left is a "better" figure in a theory of object perception proposed by Garner (1974) because it is the same shape when it is reflected around any axis or rotated into any orientation. The figure on the right is not a "good" figure because there are no axes in which the shape replicates itself over rotation or reflection.

a shape (good shapes are identified and categorized better than bad ones) could be predicted from the number of rotations and reflections ("R & R subsets," as he called them) a shape could undergo and still be perceived as the same shape (i.e., how much reflectional symmetry the shape had). For instance, a circle retains its shape under both rotation and reflection over more transformations than a square or complex figure such as a multiangled polygon (see Figure 2.29).

More recently, the role reflectional symmetry might play in object recognition has been a major source of debate centered on whether objects are represented in perception and memory by their spatially invariant volumetric parts or by multiple views (Biederman & Gerhardstein, 1993; Tarr & Bulthoff, 1995). It is not very important for the present purposes to fully understand this debate (see Palmer, 1999, for a thorough overview), but a major part of it concerns changes in performance when objects are either rotated or reflected (i.e., change their orientation or sense of direction). Some investigators report that changes in orientation and reflection do not influence object recognition (Biederman & Cooper, 1991), while others report that they do (Edelman & Bulthoff, 1992).

These debates have taken place in the context of trying to determine how to define objects and what their basic parts might be. But from a reference frame perspective, the value of manipulating certain types of transformations such as reflection is different from when one is focused on whether reflection disrupts object identification or not. Nevertheless, at least one of the basic questions is the same. How does a visual stimulus (whether object, part, scene, etc.) maintain its unity under spatial transformations? A hierarchy of spatial reference frames is one way in which this could be achieved.

In a study published some years ago (Robertson, 1995), I asked subjects to judge whether a letter was normal or mirror reflected and I measured how fast they could make these decisions under different spatial transformations. The experiment was designed to examine how spatial frames might influence performance when letters were shown in different halves of the visual field. Performance asymmetries between right and left visual fields are often attributed to differences in hemispheric function, and I wanted to see whether reference frames could account for left/right differences by rotating the stimuli so they were aligned with the midline of the body. By chance, the design included reflectional symmetry both around fixation and around the letters themselves. The relevance of the findings to hemispheric differences can be found in the original paper and indeed performance asymmetries followed the frame rotation. For the purpose of discussing sense of direction and multiple frames, I will focus on what happened in a baseline condition where letters were presented only in the right or left visual field. Reflection was manipulated either around the letter itself or around the center of the screen (which was also where the eyes remained fixated).

The letters F, R, E, or P were presented in either their normal or mirror image reflections 4.5 degrees to the right or left of fixation, and a group of normal perceivers were instructed to report whether the letters were normal or mirror image reflections. Responses were examined as a function of the location and reflection of the letter on the previous trial (prime). For instance, response time to report that an F was normal on trial N (probe trial) was coded relative to the reflection on trial N-1 (prime trial). It was also coded as in the same or different visual field and whether it was the same or different letter.

When the reflection and location were the same (Figure 2.30a) reaction time was faster than when either the reflection or the location of the prime and probe were different (Figure 2.30b and 2.30c). But more interestingly, reaction time was just as fast (in fact, slightly faster) when both reflection and location changed (Figure 2.30d) as when neither changed (Figure 2.30a). This outcome was evident whether the letter itself changed or not. In other words, it was not the letter shape nor the reflectional symmetry around the letter itself that produced the beneficial priming effects. Rather, it was reflectional symmetry in the global frame of reference around fixation.

Rotation in 2-D plane of the page has also been shown to increase the time to identify a shape (Jolicoeur, 1985), producing mental rotation functions similar to those observed when participants are asked to make re-

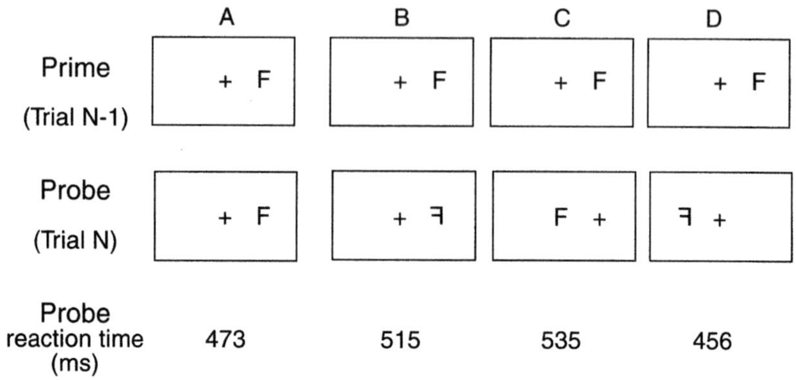

FIGURE 2.30. Examples of prime and probe pairs when both location and reflection were the same (a), only intrinsic reflection changed (b), only location changed (c), and both reflection and location changed (d). Mean response times for a group of normal perceivers are presented below. Both a and d are faster than b and c. (Adapted from Robertson, 1995)

flection judgments (Cooper & Shepard, 1973). However, in many studies when identification rather than a reflection judgment is required, there seems to be something special for a stimulus presented at 180° from upright where a flip around the horizontal axis is all that is needed to normalize it to upright. For instance, upside-down letters can produce a dip (Figure 2.31) in the normal linear mental rotation function from zero–180° degrees. Somewhat paradoxically, an upside-down letter is easier to

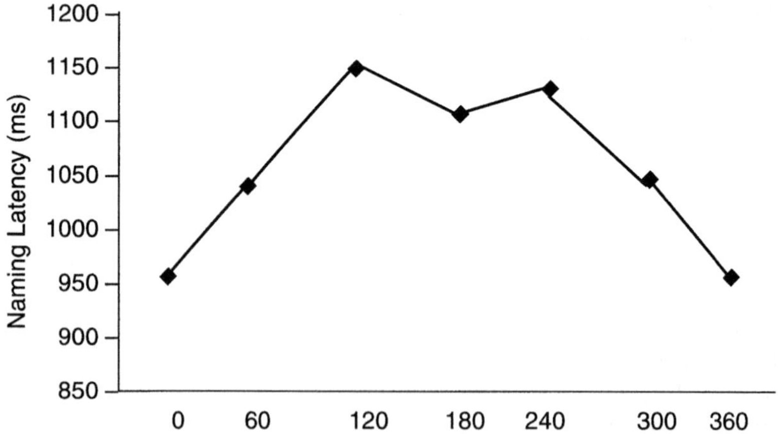

FIGURE 2.31. Mean response time to determine whether a shape is normal or reflected as a function of orientation from upright. The dip at 180° is not consistent with mental rotation around the picture plane (see text).

recognize than one presented at 120° from upright. The dip at 180° is not consistent with a smooth linear rotation around the picture plane, but rather faster identification of the shape can be made by reflection, which only requires a change in sign in a reference frame.

When considering axes and spatial transformations, reflection is simply the mirror image of a stimulus, or one of a family of symmetries that influences speed of processing (Garner, 1974; Palmer, 1999). The power of reflectional symmetry in a stimulus is undeniable. For instance, symmetrical forms are more likely to be perceived as figures in figure/ground displays (Figure 2.32). The more recent studies discussed above have demonstrated that reflectional symmetry around an axis that is not usually considered object-based but defines locations in a more global frame is also an influential factor in perception and supports the importance of sense of direction in a hierarchy of reference frames.

Neuropsychological Evidence for Reflection as an Independent Component

Perhaps the most convincing evidence that reflection is a separate component of spatial frames again comes from the neuropsychological literature.

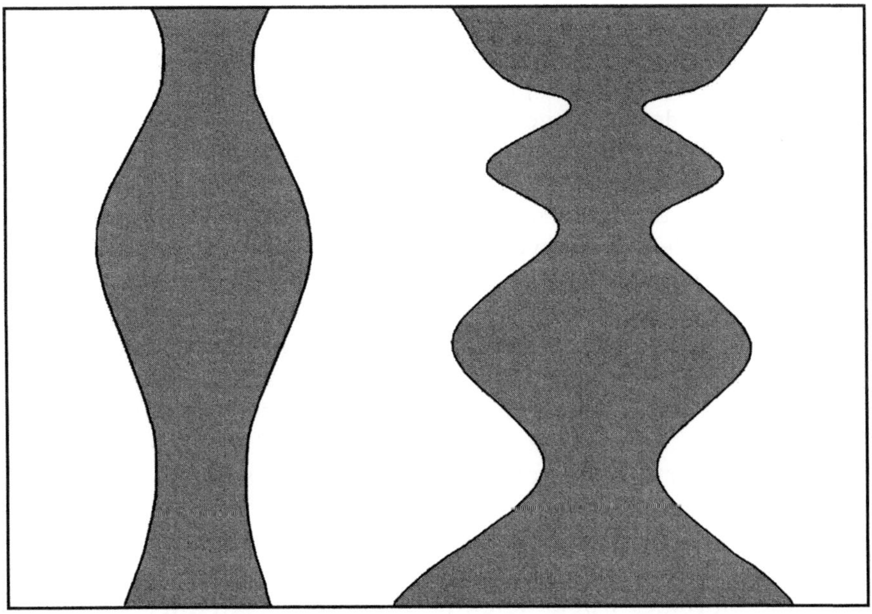

FIGURE 2.32. The symmetrical parts of this figure/ground display are more likely to be perceived as figure than ground.

A rare condition known as Gerstmann syndrome (Gerstmann, 1940) affects the ability to determine reflection or sense of direction of visual stimuli while leaving the ability to accurately report orientation and location intact. It would be of interest to know how patients with this syndrome respond to reflection around the global frame of reference, especially since the syndrome has been most often associated with left ventral lesions that may also be involved in local identification. Does this type of lesion disrupt the reflection of local frames while leaving global frames intact? Whether it does or not, Gerstmann syndrome clearly demonstrates that reflection perception can be affected without affecting other components of spatial reference frames.

A complete spatial reference frame appears to require the integration of spatial components processed by different areas of the brain. In this sense, reference frame representation is a widely distributed process that likely requires a network of activity, yet processing an individual component of a reference frame appears to be more specialized. The various ways in which object and space perception break down may not be so surprising when considering multiple spatial frames, their components, and how they influence normal perception and attention.

□ Unit Size

There is one component of spatial frames that I have left for last, because it is in some ways the most problematic, and that is the scale or unit size. All measuring devices have a base scale that defines distances. In construction-related industries, one often hears the question about whether a map is drawn "to scale," meaning that the relative distances or contours on a map correctly represent the true spatial properties. They are proportionally equivalent. Whatever unit size is adopted, each point has a one-to-one correspondence with the space being measured.

But does it work this way in perception? Our experience suggests that it does, at least to a first approximation. Perceiving the two circles in Figure 2.33 as the same shape seems simple, although their sizes are very different. If we plot them on the same reference frame as in Figure 2.34, it would be difficult to extract the equivalence of the two circles. One would be larger than the other in absolute values. But if we consider separate spatial frames, each centered on one of the circles as in Figure 2.35, then each circle is described with its own unit size. If the calculations for the diameter of each circle are performed in the global frame, then the outcomes would differ, but if they are performed within two different frames intrinsic to each circle with the frames only differing in scale, then the outcome could be the same. The shapes would then appear equivalent in shape (Palmer, 1999).

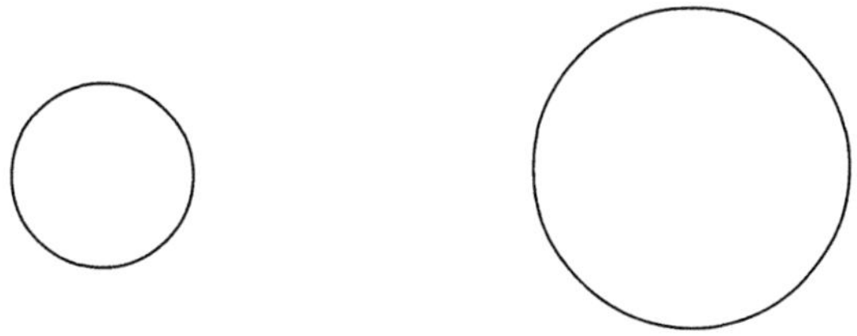

FIGURE 2.33. How does the visual system know that these two circles are the same shape but different sizes?

However, there remains a problem. We now know that the shapes are the same because the computations performed within each circle's reference frame produce the same results, but there is nothing in these results that tells us the circles are different sizes. A way to compute that the circles are different sizes is for each individual object-based reference frame to be compared in a more global coordinate system such as in Figure 2.34. This frame makes it easy to determine that the circles are in different locations and to compute their relative sizes and distances from each other. Both the global and local reference frames are required to obtain all the information we need to perceive the circles as the same shape but having different sizes.

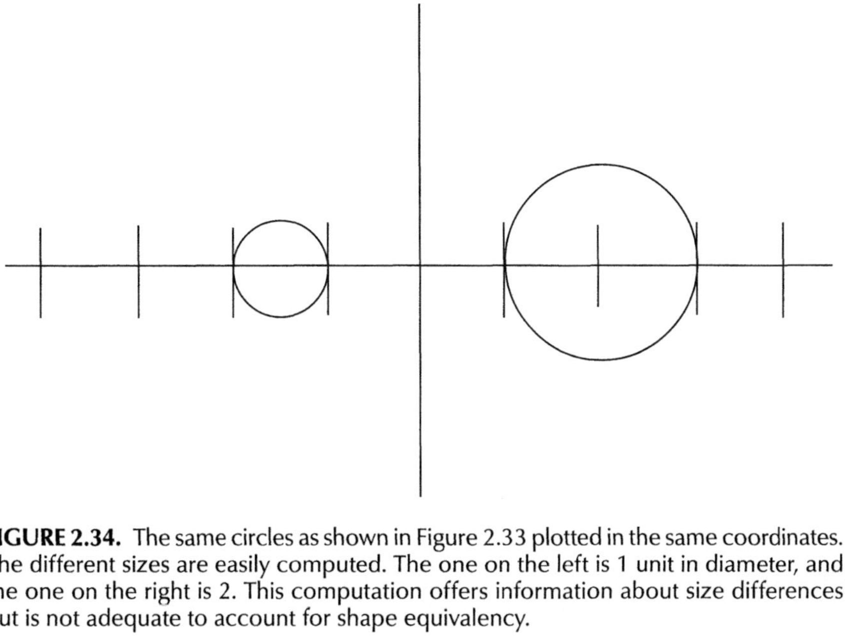

FIGURE 2.34. The same circles as shown in Figure 2.33 plotted in the same coordinates. The different sizes are easily computed. The one on the left is 1 unit in diameter, and the one on the right is 2. This computation offers information about size differences but is not adequate to account for shape equivalency.

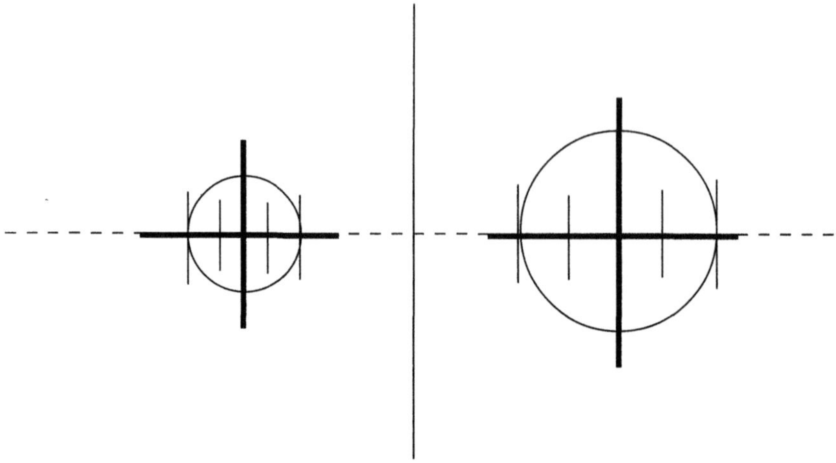

FIGURE 2.35. If each circle contained its own spatial frame with point-to-point mapping between the two, then shape equivalency is evident. The relationship to the global frame provides the information needed to know that they are different sizes.

This argument also applies to the case in Figure 2.36. There appears to be only one set of coordinates in the figure, but that is because the global and local frames are overlapping and superimposed on one another. We can conceive of a reference frame centered on each circle that would be useful in evaluating shape equivalence, and a third reference frame that would be useful in calculating the difference in size. Each may have different unit sizes. Perceiving stimuli as the same or different shapes is most efficiently derived from reference frames centered on each object that often have different unit sizes, while the relative size and location of the

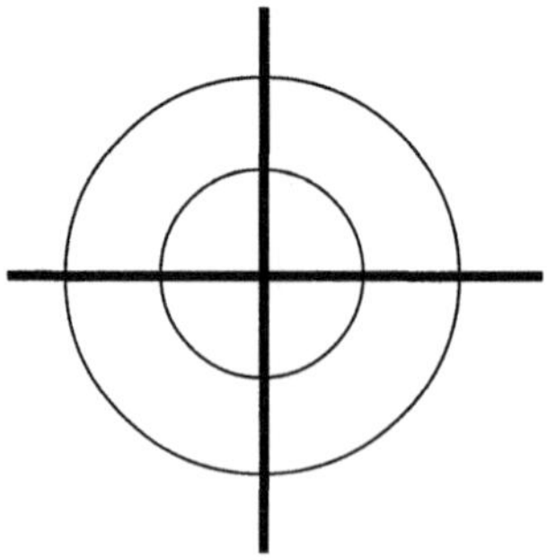

FIGURE 2.36. Example where each circle's intrinsic reference frame and the global frame overlap. Only unit size can differentiate the two.

objects is most efficiently derived from a more global frame. The internal metric of the frame intrinsic to the object as well as the metric of the global frame are necessary to calculate size and shape similarities and differences.

It is appealing to conclude that local reference frames (centered on the circles) tell us *what* items are, while global reference frames tell us *where* items are. However, this is only part of the story. Is the small circle in Figure 2.37 a hole, a nipple, or the bull's eye of a dartboard? The local frame provides information about shape by defining the relative position of each object's parts, and this shape can constrain what the objects can be. However, it is the context or relative values of visual features between different shapes as well as the combination of different features that will ultimately determine what an object is. The co-location of the background color and the small circle within the pattern on the left of Figure 2.37 signals the visual system that the small circle is more likely to be a hole than a nipple. A different color from the background, as in the picture on the right, indicates something quite different.

Although little is known about how shape equivalency is biologically achieved, we do know that brain damage affecting the ability to see where a shape is located also affects the shape's perceived size as well as what features (e.g., color, motion) are assigned to the shape (Bernstein & Robertson, 1998; Friedman-Hill, Robertson, & Treisman, 1995; Robertson, Treisman, Friedman-Hill, & Grabowecky, 1997). Binding shapes to features is disrupted with parietal lobe damage, which also affects the selection of spatial reference frames (as would be expected if there is no there there). These findings will be fully explored in chapter 6, where feature

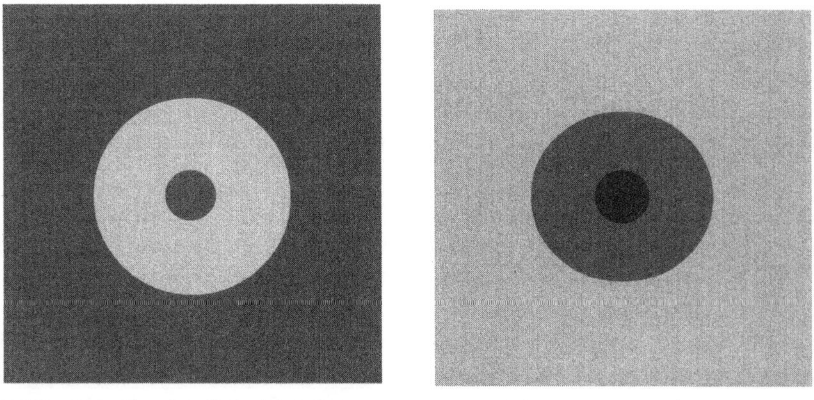

FIGURE 2.37. The small circle in the pattern on the left appears as a hole in a donut because it is the same color as the background, while the small circle on the right does not.

binding and attentional function are discussed more fully. They are mentioned here only as a reminder that spatial deficits affect more than navigation, attentional search, and spatial calculations. They also affect how objects are perceived including unit size.

☐ Summary

Together, the findings discussed in this chapter indicated the necessity of multiple spatial frames to incorporate a number of seemingly disparate results. Accounting for many perceptual phenomena seems to require the notion of spatial reference frames that unify various object/spaces in visual awareness. These frames can be defined more globally or more locally and can be linked to the retina, viewer, gravity, individual objects, or the scene as a whole. The discussions in the present chapter have focused mainly on stimulus factors that set the parameters of spatial frames in a bottom-up fashion: orientation, origin, sense, and scale. I touched briefly on the role of attention in frame structure when discussing evidence from patients with neglect and from a rare person with abnormal directions in ballistic movements. The role of top-down processing in frame selection was not a topic of the present chapter, but there is evidence that attentional control can overcome bottom-up information that enters awareness, and frame selection may play a role. When a new frame is selected, it then seems to guide spatial attention. Frame selection will be more fully explored in the next chapter.

Neuropsychological evidence has demonstrated that the components contributing to spatial reference frames can be independently effected by damage to different areas of the human cortex. The computation of space (at least the space that enters awareness) is widely distributed, while the components that create that space appear more localized. The debate should not be over whether space processing is distributed or localized. Rather, within a distributed system, there can be localization of components. Both localization and distribution are part of the dance.

3

CHAPTER

Space-Based Attention and Reference Frames

By now I hopefully have established that the fundamental components of spatial reference frames, namely orientation, origin, sense of direction, and unit size, are all factors that must be taken into account in spatial vision. All are necessary for the representation of spatial reference frames, and there is both neurobiological and cognitive evidence that they are critical for object identification and recognition as well. Although the study of reference frames in object perception has had a long history, studies of how reference frames might guide attention and/or how they are selected have had a very short one. In this chapter I will explore some of what we know about how attention selects locations, resolution, and regions of space and what role spatial reference frames might play in this process.

☐ Selecting Locations

When one speaks of space, location immediately comes to mind. Where are my keys? Where did I park the car? Where is the light switch? Where did I file that paper? A game of 20 questions may be in order to help guide us to whatever it is we are seeking. Is that manuscript at home? If yes, is it in my filing cabinet or one of the many stacks on the floor? If in one of the stacks, is it in the one with the neuropsychology papers or the one about normal vision, or perhaps the one that catches everything else? If in the "other category" pile, is it near the top or bottom? And on it goes. Where, where, where, where—down through the hierarchy of "objects" (home, stacks of paper on the office floor, topics, etc.).

I have discussed some evidence that suggests that locations in perception can be defined in selected spatial reference frames at different hierarchical levels of object/space representations. In this section I will set the hierarchical part aside for the most part and address the question of attentional selection of a location in a way that is more familiar, namely as if there is a unitary spatial field with objects in different places. Nevertheless, it should be kept in mind that attention to a location within any spatial frame that is selected could guide attention in the same way.

Perhaps because of the emphasis on spatial locations in communication, action, and everyday living, there are a large number of studies concerned with how we select a location that is of particular relevance at any given moment in time. How does attention enhance sensitivity to this location or that? Is there some mechanism that scans a display serially as eye movements do, from one location (or object) to another? Are there cases where all locations can be searched in parallel (all locations or all objects at once)? How do the visual characteristics of objects change search patterns?

We know a fair amount about the answers to each of these questions from the cognitive literature. A cue that predicts the location of a subsequent target enhances detection time for targets appearing in that location and slows detection time for targets appearing at uncued locations (Posner, 1980). Experimental evidence has confirmed that the costs and benefits can be due to modulations in sensitivity and not only to changes in response bias (Bashinski & Bacharach, 1980; Downing, 1988). Many also argue that spatially scanning a cluttered array requires a serial attentional search from one object to another or from one location to another under the right conditions (Treisman, 1988; Treisman & Sato, 1990). In the laboratory, detection rates for a predetermined target can increase linearly with the number of distractors in a display (see Figure 3.1a). Attention seems to sample each item or group of items in different locations serially (Figure 3.1b). Other work has shown that this type of scan can be guided in particular ways by prior encoding, such as grouping or differential weighting of basic visual features (Wolfe, 1994). These processes can reduce the slopes of the search functions and also the 2:1 ration between slopes when the target is present versus when it is absent.

On the other hand, unique features in a cluttered array (Figure 3.2a) do not require spatial attentional search, but instead "pop out" automatically (Treisman & Gelade, 1980). In this case, detection rates do not increase linearly with the number of distractors in the display (Figure 3.2b).

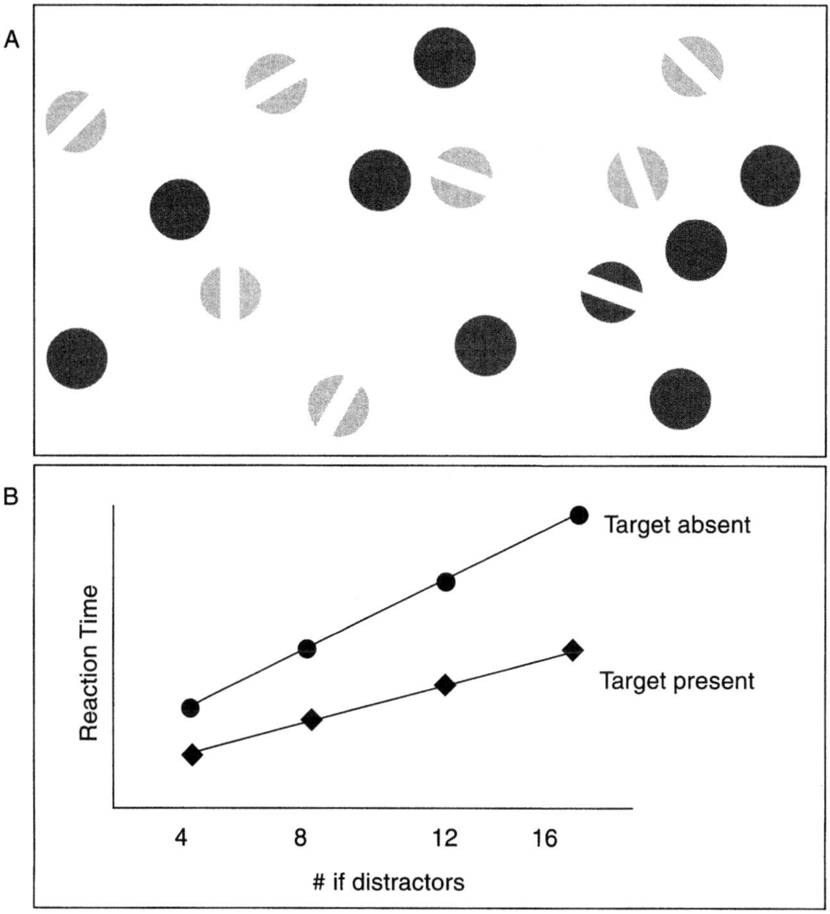

FIGURE 3.1. When searching for a target that is a red (dark gray) dot with a line through it among distractors that are either solid red dots or blue (light gray) dots with lines through them (a), response time increases linearly as the number of distractors increases (b). On average, more distractors would have to be searched to determine if a target is absent than to determine if it is present, producing an interaction between number of distractors and target presence. Note that (a) is a conjunction search display because the target is the conjunction of the features in the distractors and that the colors were closer in luminance. (Adapted from Eglin et al., 1989.)

Spatial information is needed for serial but not for feature search. Consistently, severe spatial deficits do not affect pop out, but they do affect serial search (see chapter 5).

We also know something about the functional pathways in the brain that select location (see Figure 1.9). A cortical network associated with a

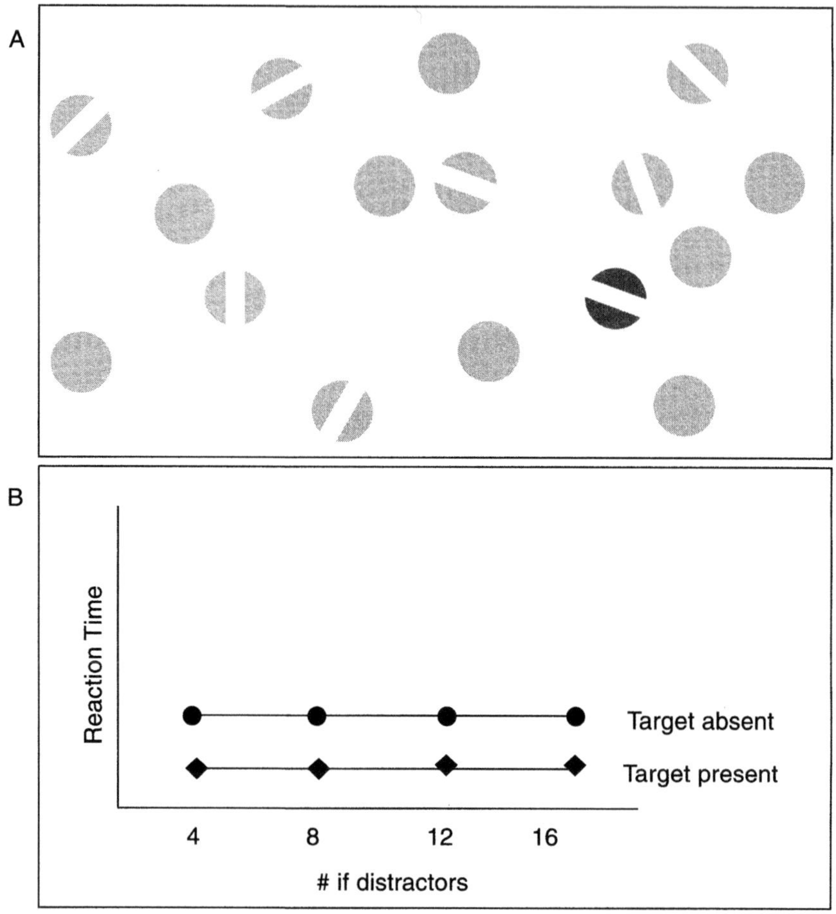

FIGURE 3.2. When searching for the same target as in Figure 3.1 but now with both the solid dots and the dots with lines through them being blue (a), response time does not differ as a function of number of distractors (b), and the interaction between target presence and number of distractors disappears. Note that (a) is a feature search display because the target contains a unique feature (in this case the color red) that is not in any of the distractors.

dorsal processing stream (the dorsal occipital-parietal-frontal cortex) seems to direct attention to selected locations (Posner, 1980). Attention must then be disengaged to move to another location when needed. Damage to the parietal lobe of this stream disrupts the ability to move attention to new locations (Posner, Walker, Friedrich, & Rafal, 1984). Consistently, parietal lobe damage also disrupts the ability to move spatial attention through a cluttered array (Eglin et al., 1989) but not to detect the presence or absence of a unique feature that pops out (Estermann, McGlinchey-

Berroth & Millberg, 2000; Robertson et al., 1997). Ventral cortical areas that are believed to encode object features (e.g., color, shape, brightness, etc.) are sufficient to see a target pop out but not to guide attention to search for one that does not.

In addition, areas of the frontal lobe abutting the frontal eye field (supplementary eye field) seem to be involved in maintaining the spatial location of a target in memory (Goldman-Rakic, 1987). The frontal eye field is also involved in oculomotor programming that accompanies (often follows) attentional movement to a location (Henderson, Pollatsek, & Raynor, 1989; Posner, 1980). Spatial attention and eye movements are generally linked in the normal brain (Corbetta et al., 1998), and this makes good sense. Attention to detail is more efficient when visual information falls on a region of the eye engulfing about 2.5° at fixation (i.e., the fovea). An eye movement may pull attention with it, or attentional selection may pull an eye movement with it under normal everyday conditions. However, in the laboratory, eye movements and spatial attention have been successfully dissociated. Attention can clearly be where fixation is not.

Eye movements and attention are also closely linked within parietal cortex (Corbetta et al., 1998). However, some other mechanism signals many eye movement cells within this area, as they begin to fire in anticipation of an eye movement to a targeted location (see Colby & Goldberg, 1999; Andersen, Batista, Snyder, Buneo, & Cohen, 2000). Here, attention seems to precede movement.

Also, another part of this system (the cingulate gyrus) interacts with frontal and parietal areas and may provide motivation to attend as well as to perform accurately (Mesulam, 1985). Even a small reduction in motivation can erase the will to move attention to areas outside the present line of sight.

Finally, there are hemispheric differences that are fairly robust, at least in humans. Damage to posterior areas of the dorsal pathway are more likely to cause spatial deficits when in the right hemisphere, while damage to posterior areas of the ventral system are more likely to cause language deficits when in the left hemisphere. The nature of these deficits has been discussed extensively under a separate cover (Ivry & Robertson, 1998).

☐ Reference Frames and Spatial Selection in Healthy and Neurologic Patient Populations

Many studies of spatial attention have placed stimuli around the center of fixation in order to control for such factors as eccentricity, side of cue, hemisphere directly accessed, and so forth, with little thought of inherent spatial biases. Yet one spatial bias that keeps appearing in the attentional literature is a rightward one (e.g., Drain & Reuter-Lorenz, 1996). Most investigators tend to ignore this bias and find it something of a nuisance. It is often left hanging because it is unexpected and has little relevance for the question the experiments were designed to answer. When investigators have paid attention to this bias, they have mostly been concerned with differences that could reflect functional hemispheric asymmetries. For instance, Kinsbourne (1970) suggested that the rightward bias observed in normal perceives reflected a vector of attention toward the right due to increased activation or arousal of the left hemisphere by ubiquitous language processing in humans.

Neurobiological evidence suggests that this bias in attention occurs through cortical/subcortical interactions between the two sides of the brain (see Kinsboure, 1987 or Robertson and Rafal, 2000, for details). Initial evidence was derived from animal research, which showed that a unilateral posterior cortical lesion produced neglect-like behavior (a right hemisphere lesion made the animal orient toward the right). However, when the superior colliculus on the opposite side was ablated in the same animals, the rightward orienting disappeared (Sprague, 1966). We also know from the literature in human neuropsychology that a lesion in the right parietal lobe can cause left neglect, but that symmetrical lesions in both parietal lobes do not produce a spatial bias, instead bringing attention back to the center (Balint, 1909, Holmes & Horax, 1919).

These observations can be explained by a functional cortical/midbrain loop like that represented in Figure 3.3. The superior colliculi (SC) are mutually inhibitory, with activation levels modulated by frontal and parietal connections. This architecture could explain the rightward bias as stronger inhibition of the right SC, by the left SC which would arise from stronger activation of frontal-parietal areas in the left hemisphere (the right being in charge of moving attention to the left, and the left to the right). In other words, anything that produces a hemispheric imbalance of cortical activation of frontal-parietal areas (stroke being the most dramatic) would change attentional biases (see Kinsbourne, 1970, for a proposed theory of attentional vectors). Kinsbourne argued that the left hemisphere's role in language processing would produce higher levels of overall activation in that hemisphere in normal perceivers. This in turn would produce more activation of the left SC, and due to its inhibitory effect on the right SC, this would decrease the normal right SC's inhibition on the left, resulting in a vector of attention biased toward the right. Given the predominance of language functions in the left hemisphere in the general population, the result would be a population bias of attention to the right. The degree of this bias in each individual would depend on the balance between activation and inhibition within this cortical/SC network.

FIGURE 3.3. The superior colluculi are mutually inhibitory but receive excitatory input from parietal and frontal cortex.

Kinsbourne (1987) went on to argue that unilateral neglect observed more often with right hemisphere than left hemisphere damage was a consequence of disrupting the overall normal balance between the hemispheres with its slight rightward shift. When the right parietal lobe was damaged, activation of the right SC would be significantly reduced, and this in turn would reduce the amount of inhibition on the left SC that was normally present. The consequent increased activation in the left SC (from the intact left parietal input) would increase the rightward bias. The result of a cortical lesion in the right hemisphere would then be a dramatic swing of attention to the right side, which is exactly what happens.

This is a simplified account of the functional neuroanatomy that has been offered to explain the rightward attentional bias that is often reported in the cognitive literature. Why this directional bias exists at all is unclear, although attempts to relate it to other functional asymmetries such as language have been attempted. In reality, the rightward bias is no more or less puzzling than the population bias for right-handedness, and the rightward attentional bias appears often enough to conclude that it is a real phenomenon.

Although population spatial biases are interesting in their own right, it is not the question I am concerned with here. Nevertheless some discussion of why it might be present seemed warranted because later in this chapter I will introduce studies that have exploited this rightward bias, using it as a marker to study attentional allocation and spatial frames of reference.

Reference Frames Guide Location Selection in Normal Perceivers

Some years ago Marvin Lamb and I wondered whether the rightward spatial bias would only occur within a viewer-centered reference frame (right vs. left visual field) or would also occur in other reference frames (Robertson & Lamb, 1988, 1989). At the time there was great concern about why some visual field differences in performance (which were presumed to reflect functional hemispheric differences) were so difficult to replicate. Although lexical decision tasks could usually be relied on to produce a right visual field advantage (left hemisphere), single letters, different types of objects, pictures, scenes, colors, etc. were far more variable and produced a great deal of head scratching. Some researchers argued that attentional allocation or variable strategies changed the hemispheric balance in ways that often were not predictable (Morais & Bertelson, 1975). When the subject's ability to volitionally allocate attention was controlled, the data became less variable. In addition, some spatial biases to the right visual field were common enough to make researchers wonder whether these were due to the hemisphere of input or to other types of processing mechanisms such as those that guide spatial attention (see Efron, 1990).

FIGURE 3.4. Examples of normal and reflected letters used by Robertson and Lamb (1988, 1989).

We approached the question by varying the orientation of stimuli around fixation in such a way that a spatial coordinate was defined that changed right and left relative to the viewer but was maintained relative to the stimulus. In the first experiment we showed letters in either the left or right visual field in a manner typical of human laterality studies used with normal perceivers. Letters were flashed about 3.5° from fixation for 100 ms (too fast to make saccadic eye movements), and subjects were told to keep their eyes fixated on a central plus sign at all times. The letters were presented in either their normal or mirror image reflection (Figure 3.4), and subjects simply responded whether the letters were normal or reflected as rapidly as possible. We adopted this particular manipulation because we could control for the distance between fixation and any critical features of the letters that might change response time (see Figure 3.5), such as how close the most informative features were to fixation. For instance, an E's three points would be closer to fixation when it was normal and presented in the left visual field and when it was reflected and presented in the right visual field, while the three points would be farther from fixation when it was reflected and presented in the left visual field and when it was normal and presented in the right visual field. If a rightward advantage was still observed under these conditions where the eccentricity of stimulus features were counterbalanced over trials, then it would be difficult to attribute the effect to visual feature analysis.

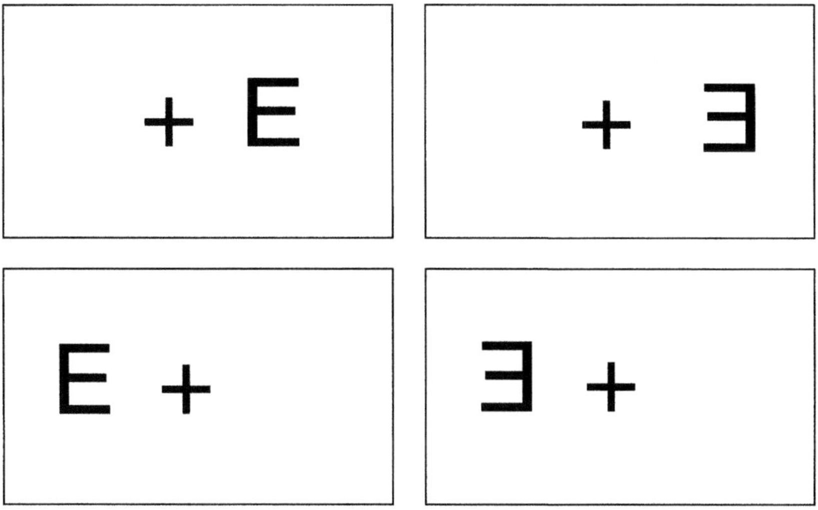

FIGURE 3.5. Example of one of the letters and its reflection and location variations used in the studies by Robertson and Lamb (1988, 1989) presented on the right or left side of fixation (represented by the +).

We found a robust rightward advantage for all stimuli (Figure 3.6), but the real question was whether this rightward advantage would be maintained when they were presented rotated 90° from upright but in the upper or lower visual field, and it was. We made sure this could not be attributed to head tilt or rotation of the participants themselves by using a chin rest and head restraint that kept their heads upright at all times. They were reminded to fixate on the central plus sign throughout the block of trials and to respond to the letters' reflections as if they were upright. In one block, the letters were oriented 90° clockwise from upright, and in another block they were oriented 90° counterclockwise (Figure 3.6). But now the stimuli appeared in the upper or lower visual field, again about 3.5° from fixation. We can think of the letter's orientation as defining the top of a reference frame either pointing leftward or rightward relative to the viewer. The right side in the frame thus became the upper location on the screen when the stimuli were rotated counterclockwise but the lower location on the screen when the stimuli were rotated clockwise.

The most striking result was that the rightward bias within the frame was present in both rotated conditions. There was a lower visual field

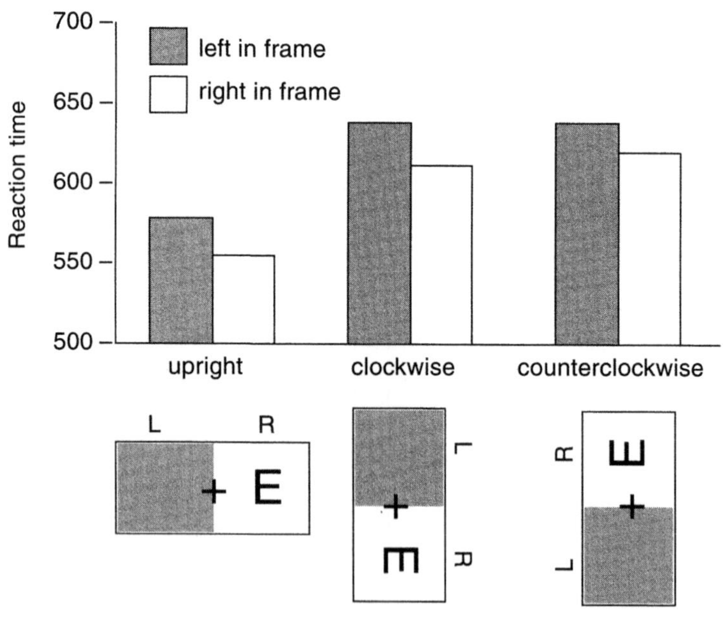

FIGURE 3.6. Mean response time to respond whether letters were normal or mirror reflections as a function of stimulus field and frame orientation. (Adapted from Robertson and Lamb, 1988.)

advantage when stimuli were presented 90° clockwise and an upper visual field advantage when they were presented 90° counterclockwise. Within display-centered reference frames with an origin at fixation, these were both on the right. Note that when letters were presented upright, it was impossible to determine whether the rightward bias was due to the position in environmental, viewer, retinal, or display coordinates. Given the results observed in the rotated conditions, we can conclude that right and left locations were defined relative to display-based coordinates.

Note that the reference frame was not centered on the target itself. Left and right were instead defined as locations in the reference frame through an axis with its origin at fixation. Left and right locations where the stimuli could appear were defined relative to this origin. The orientation of the letters defined the sense of direction of the frame, but attention appeared to select the frame that moved with the orientation. This frame is not object-based in the traditional sense (Humphreys, 1983; Palmer, 1989; Quinlan, 1995; Rock, 1990, etc.) because the origin was not centered on the object. It was centered at fixation or what could be thought of as the statistical center of the entire block of trials. Because of this distinction I will refer to the frame as a scene-based frame.

In a follow-up study I used a priming method to determine whether stimulus orientation had to be blocked in order to observe such results (Robertson, 1995). Did subjects only adopt a scene-based frame when a series of stimuli all appeared in the same orientation or would subjects adopt frames more transiently?

In this study a prime (letters) was presented at fixation on every trial, and it was randomly oriented either upright, 90° clockwise, or 90° counterclockwise (top row of Figure 3.7). This prime informed the participants that the upcoming letters in the periphery would be oriented in the same way as the prime but it did not inform them where the target letter would appear. The peripheral target letters were again either normal or reflected, and the prime was also either normal or reflected, but the prime's reflection had no predictive value (it was orthogonally varied with the reflection of the target).

The results confirmed the reference frame effects we found in the blocked design. When the prime was upright, there was a right visual field advantage. When the prime was 90° counterclockwise, there was an upper visual field advantage, and when the prime was 90° clockwise there was a lower visual field advantage (bottom of 3.7). These effects were present whether the prime and target were the same or different letters, which is consistent with spatial frames rather than stimulus shape as the critical factor in producing the results.

The two experiments I've discussed so far confirm that processing speed for items on the right in a scene-based reference frame are faster than for

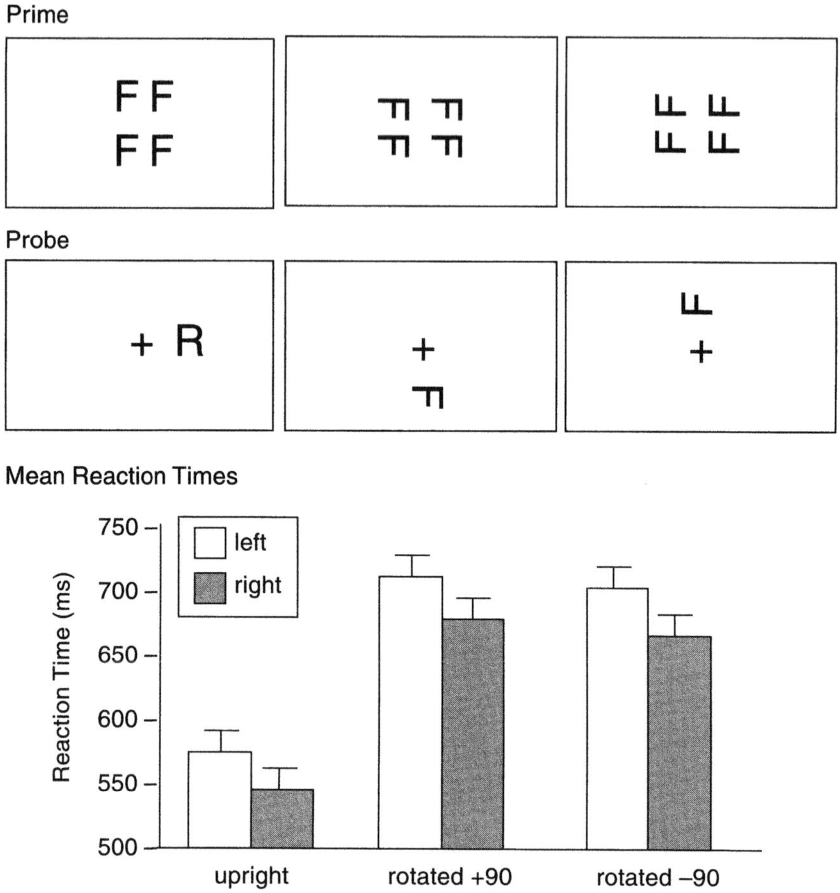

FIGURE 3.7. Example of a centrally presented prime presented either upright, 90° clockwise, or 90° counterclockwise, and a subsequent probe presented in the same orientation as the prime but off to the left or right in a scene-based reference frame centered on fixation (top). Mean response time to determine whether the probe was normal or reflected as a function of left or right side defined in scene-based coordinates (bottom).

items on the left when there is nothing in the experimental design to bias attention one way or the other. The visual placement of features of the stimuli were also controlled through varying reflection so that participants would not be encouraged to shift attention toward one side or the other by stimulus features such as the direction the letter faced. However, this does not necessarily mean that there was a rightward bias of attention per se. A population rightward shift in attention may very well explain the results, but attention was not manipulated in this experiment, and other explanations are possible without reference to attentional mechanisms (e.g., stronger weighting of a direction within a frame during perceptual organization).

To directly investigate the role of attention in producing the rightward bias, and more specifically to investigate attention's link to spatial reference frames, Dell Rhodes and I designed a series of studies in which we manipulated attention with traditional attentional cuing measures (Rhodes & Robertson, 2002). First, we changed the orientation prime that I used

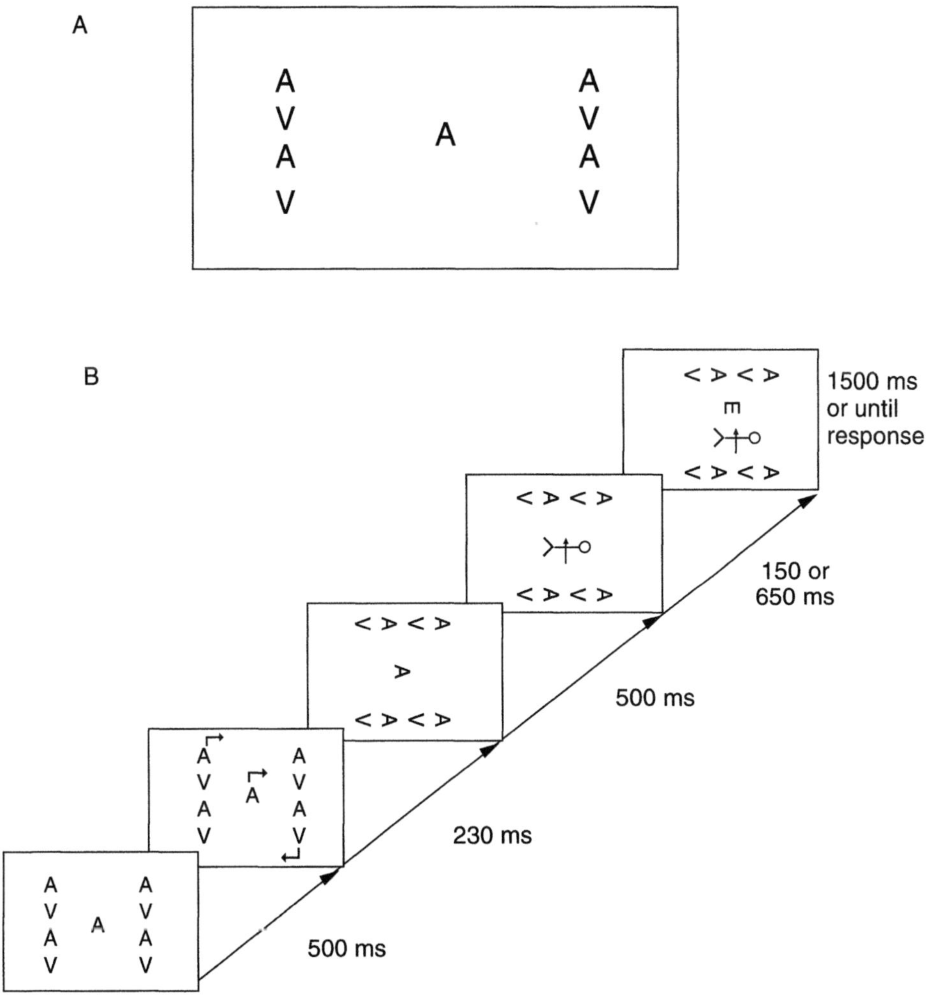

FIGURE 3.8. Example of display used as an orientation prime by Rhodes and Robertson, 2002 (a) . A trial sequence showing timing parameters, a rotation, the cue, and the target (b).

(Robertson, 1995) into a configuration of A's and V's (Figure 3.8a) to give a strong impression of a frame. Unlike in the previous experiment, this display required no response. On each trial the entire display appeared upright and either remained that way or rotated 90° in full view of the subject. As before, rotation was either clockwise or counterclockwise. Subjects were instructed to keep their eyes on the central A, since it would change into an "arrowman" figure as soon as the frame stopped rotating ("arrowman" became "arrowperson when someone questioned our terminology at a meeting at CSAIL where these results were first presented). The arrow in arrowperson was a cue that predicted where a target would most likely appear (Figure 3.8b). As before, the targets were normal or mirror image–reflected letters, appearing in the same orientation as the frame but offset right or left from fixation relative to the frame. They were presented for 100 ms, too rapid for a saccade.

As would be expected from the attention literature, responses were faster when the target appeared in a cued location (valid) than when it appeared in an uncued location (invalid). In the valid condition (when the target was where the subject expected it to be) responses were faster for targets on the right than on the left side of the frame. However, when the target appeared in the unexpected position (invalid condition), responses were slower for targets on the right than on the left. More importantly, this pattern was consistent across the different frames (Figure 3.9). It was not the absolute location in which the target appeared but its location in the frame that produced the different pattern of response time for valid and invalid trials.

Although this pattern was strong evidence for attentional processes taking place within scene-based reference frames, the difference in the pattern for valid and invalid trials was somewhat puzzling. Why were right-sided targets easier to discriminate when they were in the valid location and harder when they were in the invalid location? Further studies determined that this was due, at least in part, to conditions when arrowperson (the cue) pointed left. When arrowperson pointed to the left, the right side of space suffered. The expectation of a left-sided target appeared to require more processing resources, reducing resources at the other location—in this case, reducing resources for the right side. Again, this was the case in all three frames, supporting the importance of spatial frames in the allocation of attention. In other studies in the series we were able to factor out effects due to stimulus–response compatibility (often referred to as the Simon effect) and the baseline rightward bias, but in all cases the directional biases rotated with the frame.

Logan (1995) also studied attentional allocation in selected reference frames in a series of experiments with young college students. Instead of exploiting the right-sided bias as we did, he used a well-documented dominance of vertical over horizontal axes (Palmer & Hemenway, 1978). Stimuli

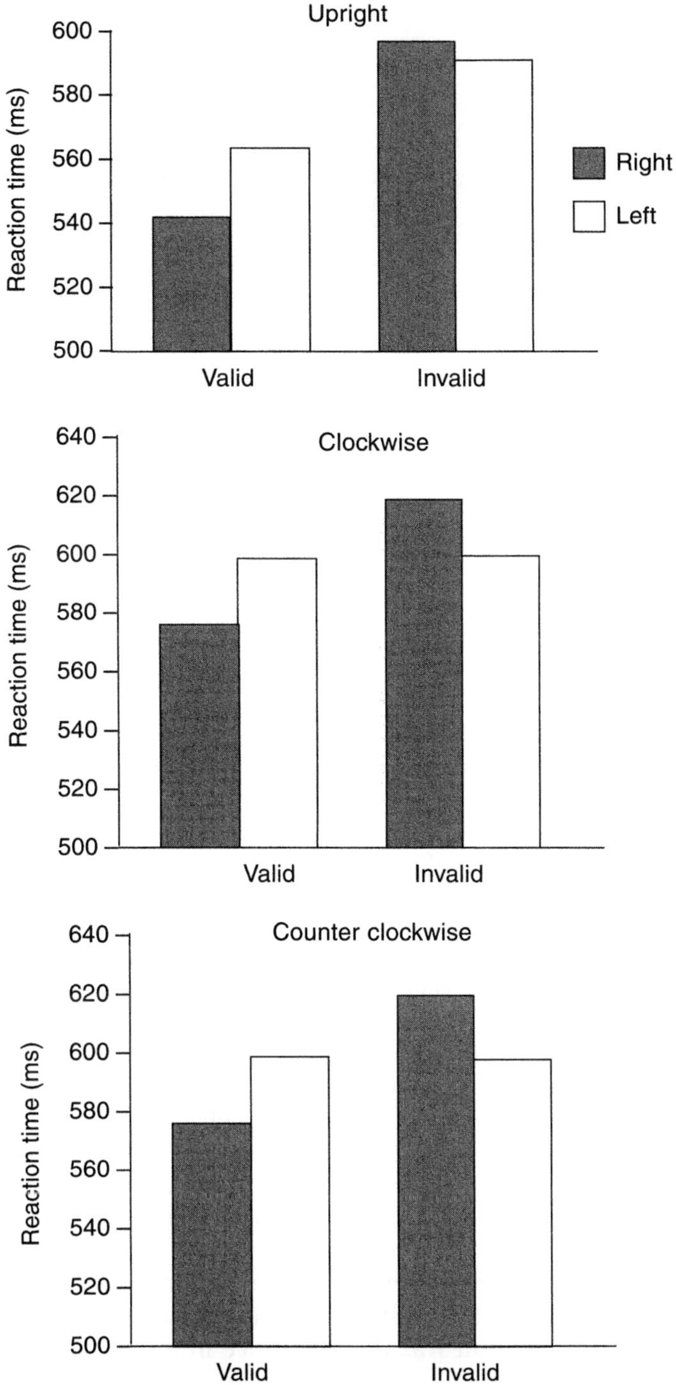

FIGURE 3.9. Mean reaction time to determine whether target letters (see Figure 3.8b) were normal or reflected for validly and invalidly cued locations under the 3 rotation conditions described in the text. (Adapted from Rhodes & Robertson, 2003.)

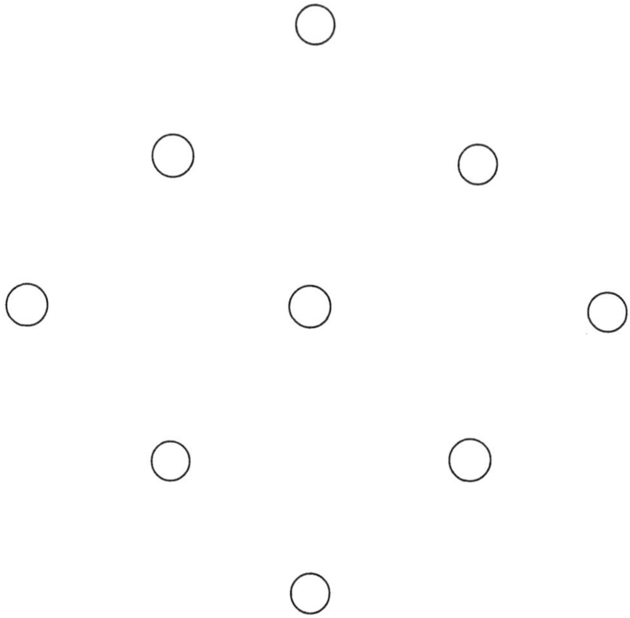

FIGURE 3.10. Representation of the 9-dot display used by Logan (1995). (Adapted from Logan, 1995.)

presented along vertical axes are responded to faster than those presented along horizontal axes.

Rather than dissociating the viewer frame from the display frame through rotation as we did, Logan (1995) dissociated fixation of attention and eyes in an upright frame. He first cued subjects to a group of 4 dots in a 9-dot display (Figure 3.10) while making sure they maintained fixation on the central dot. The 4 dots that were cued formed a diamond to the right (Figure 3.11a), left (Figure 3.11b), top (Figure 3.11c), or bottom (Figure 3.11d) of fixation. The target (a red or green circle) always appeared in one of the 4 locations within the cued diamond and subjects responded as rapidly as possible whether it was red or green.

First as expected, when performance was collapsed over the 4-dot cluster that was cued, discriminating targets positioned on the vertical axis (of the 9-dot display in viewer-centered coordinates) was 112 ms faster than discriminating targets along the horizontal axis (the 3 dots along the y axis vs. the 3 dots along the x axis in Figure 3.12). This was consistent with the vertical bias reported in the perception literature (Palmer & Hemenway, 1978). But the most impressive evidence for the role of reference frames on attention was the difference in discrimination time for

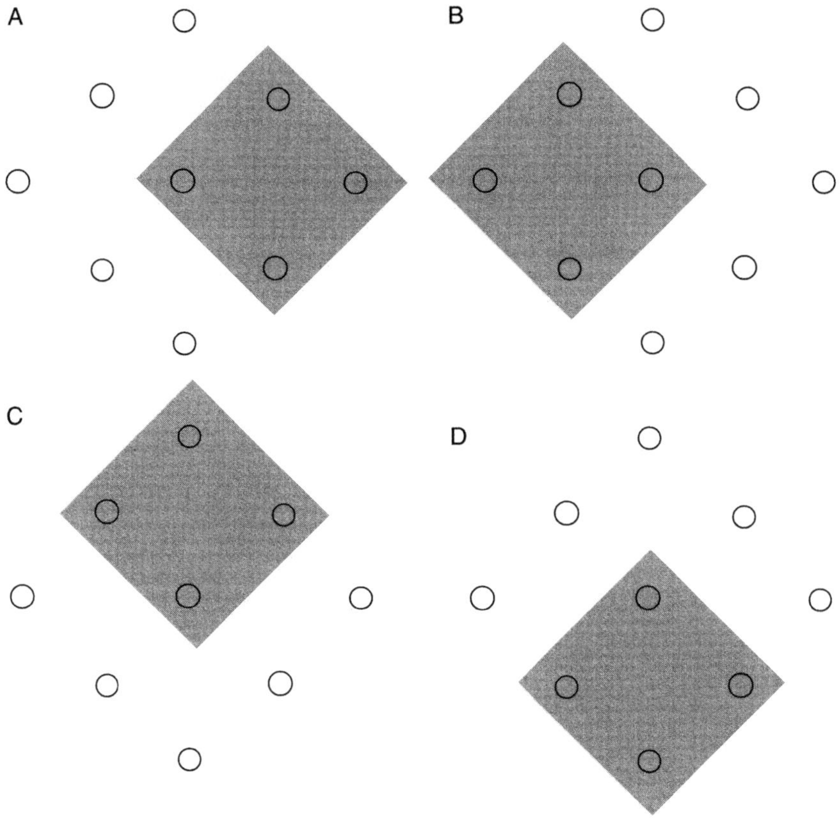

FIGURE 3.11. The 4 dot elements that were cued within the Logan (1995) study are represented in gray. Notice that the central dot of the 9-dot display was to the left (a) or right (b) within the cued region (horizontal) or to the bottom (c) or top (d) of the cued region (vertical).

targets that appeared at fixation (the central dot in the overall display). When this dot was either the lower or upper item in the cued diamond, respectively (Figure 3.11c and 3.11d), discrimination time was 126 ms faster than when the same dot was the left or right item in the cued diamond, respectively (Figure 3.11a and 3.11b). In other words, when its position was defined along the vertical axis of the cued diamond, response times were faster than when it was defined along the horizontal axis of the cued diamond. This dot never moved. It was always at fixation, but its position within a selected reference frame did change.

In another set of studies Logan (1996) addressed the question of top-down or executive control of reference frame alignment. As mentioned

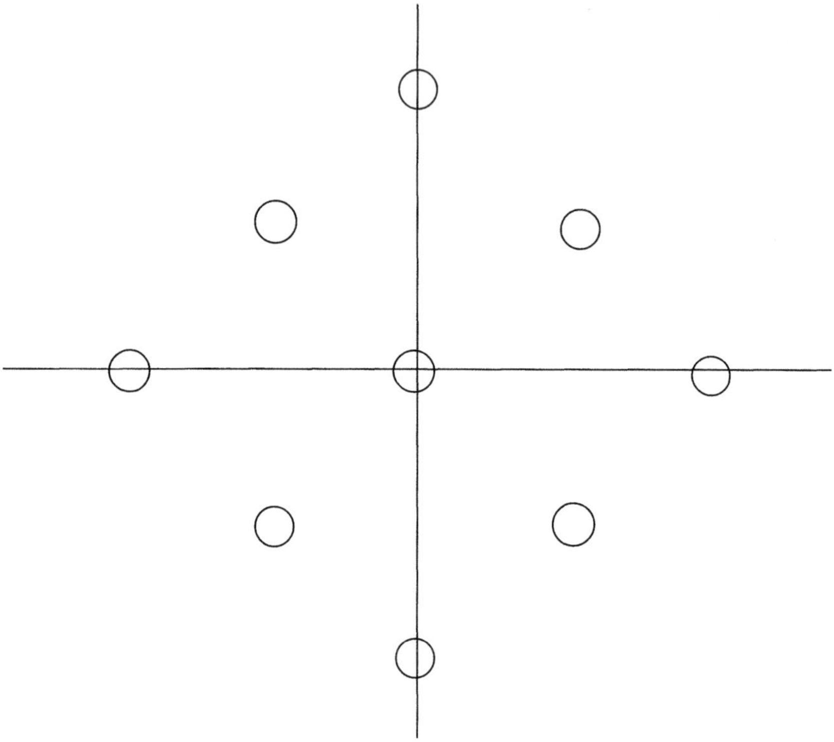

FIGURE 3.12. Horizontal and vertical locations included in the analysis of overall vertical versus horizontal response time.

in chapter 2, both elongation and symmetry can influence the positioning of reference frames (Palmer, 1980). Axes tend to be aligned with the elongated axis and symmetry of a stimulus. However, the influence of these attributes can be overcome almost entirely by executive control.

Logan presented subjects with faces where the shape of the outer boundaries of the face was elongated and could disrupt the symmetry of the face (middle pattern of Figure 3.13). On every trial he cued subjects to report the color of a dot that appeared about 1 second after the face and was either above, below, left, or right of the face. The faces were presented upright or rotated 90° or 180° from upright to dissociate them from viewer-centered frames. Neither elongation nor symmetry had much of an effect on reaction time. The major contribution was from the orientation as defined by the features of the face and the expectation of the subject. Subjects were able to all but ignore the bottom-up information that would normally contribute to reference frame alignment.

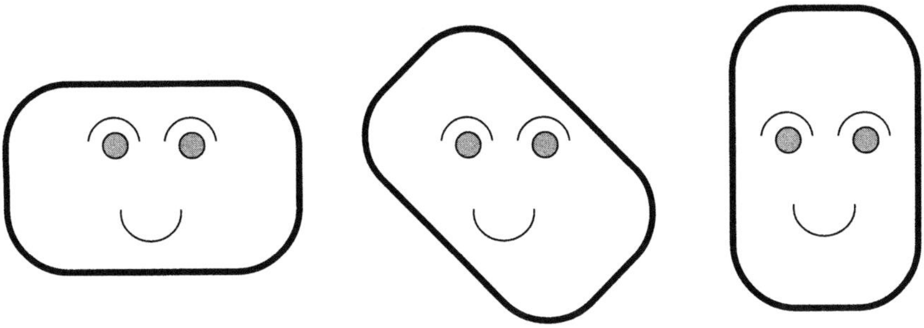

FIGURE 3.13. Example of face-like stimuli used by Logan (1996).

Some Comments on Hemispheric Laterality and Visual Field Effects

The evidence for reference frames and attention may be used to argue against the use of visual half field presentation to study hemispheric laterality in normal perceivers, but this would be a mistake. The question of the role attention can play in producing visual field differences has had a long and colorful history in the debate over the use of such methods to study how the hemispheres may contribute differently to cognition. If attention can be distributed in such flexible ways, how can we know when a visual field difference represents differences in hemisphere function and when it is the product of flexibility in allocating attention or other processing resources within a selected frame? Certain properties of stimuli might be more "attention grabbing" than others (e.g., the typically rightward-facing letters of our alphabet). Reading habits might direct more attention to the right than the left. Perhaps something in the testing environment that the experimenter did not notice could attract more attention to one side or another (e.g., a stronger light source coming from the left than the right or the monitor being closer to a right wall).

Differences in the allocation of attention have been considered for some time, and careful researchers interested in testing hemispheric differences in normal perceivers have often gone to great lengths to control the environment so as not to inadvertently attract attention to one side or the other. The assumption is that if attention is controlled so that it remains in the center, then any differences between performance for stimuli presented in the right versus left visual field can be attributed to the information directly accessing the contralateral hemisphere (stimuli presented on the right are directly projected to the left hemisphere, and stimuli presented on the left are directly projected to the right hemisphere). The results discussed in the previous section do nothing to alter the concern about attentional factors, but they do demonstrate a way in which direct access models of hemispheric differences can be evaluated for any given set of stimuli. If an upright difference rotates completely with the rotation of the stimuli, then it does not support any simple model of direct access to account for visual field differences (see Hellige, Cowin, Eng, & Sergent, 1991, for an exemption in a lexical decision task).

It is, of course, still possible (and maybe even likely) that the differences in performance that are maintained over rotation originate in initial primary cortical spaces as represented by the two hemispheres, with the left hemisphere coding the left space and the right hemisphere coding the right space relative to fixation. There must be a representation of space on which to hang the descriptors of left and right, and it might be the left hemisphere that defines the right side of rotated spatial frames and the right hemisphere that defines the left side. This would occur in more abstract computational terms. If future neurobiological evidence supports this position, then direct access models would not be entirely discredited. Perhaps feedback pathways from areas such as the parietal lobe to primary visual cortex support transformation of the early spatial encoding into a more spatially invariant spatial frame. In this way the space that is directly accessed by stimuli within the left or right visual field may form the initial basis for spatial frames that operate in extra-retinal spatial maps and provide spatial constancy when the stimuli are rotated. The left hemisphere may continue to represent the right side and the right hemisphere continue to represent the left side of the frame, but in a space that has now gone beyond retinal coordinates and visual fields. The same arguments hold for upper and lower fields.

Reference Frames and Location Selection in Neurological Patients

Sometimes the simplest of bedside tests can be as revealing as controlled tests in the laboratory. For instance, a very common bedside test of neglect is to wriggle a finger on the examiners left or right hand or on both hands together and ask the patient to deterimine when one or two fingers move. Often the patient must be reminded to keep looking at the examiner's nose because they are very likely to move their eyes in the direction of the finger movement when they see it. However, a person with left neglect will neither report the finger that wriggles on his/her left side nor tend to look in that direction, while the finger on the patient's right side appears to attract attention, and eye movements follow (unless the patient is otherwise reminded to keep them from moving).

Although patients who exhibit this response profile may in fact have unilateral neglect, they may also have a primary visual scotoma or homonymous hemianopsia (a field cut produced by an affected occipital lobe or a lesion sufficiently ventral to affect white matter projections of visual sensory information via the optical radiations). A patient with a left field cut and no neglect knows that the left side of space is present but cannot see the information presented there. Patients with field cuts will compensate by moving their eyes in the direction of the blind field in order to see information on that side. A patient with neglect will not, whether a field cut is present or not. Nevertheless, it remains difficult to determine behaviorally when a person has a field cut and neglect as opposed to neglect alone.

For a patient with left neglect who shows no sign of a field cut, another clinical exercise can be revealing. If the examiner bends his or her body through 90° so that the hands are extended vertically and aligned with the patient's body midline, neglect may be found within this new frame of reference. (I'll call this the Martinez variant because I used it to show a frame effect to my clinical colleagues at the Veterans Hospital in Martinez for the first time in 1983.) If the patient still neglects the right finger (on the patient's left side in the frame defined by the orientation of the examiner's head) and does not make an eye movement toward that side, then neglect can be documented where no field cut would be present (e.g., the upper visual field when bending rightward and the lower visual field when bending leftward). This has the potential to help resolve at least some questions that neuropsychologists must deal with about whether neglect and a visual field cut are present.

Unilateral extinction is a much less problematic spatial deficit that is a cousin to neglect (and what some consider a milder form of neglect). Patients with extinction are able to detect a stimulus on either the right or the left side of space when it is presented alone but will "extinguish" (i.e., neglect) the contralesional stimulus when items are simultaneously presented on both the right and left sides. In the Martinez test a patient with right hemisphere damage resulting in extinction would correctly report seeing the right or left finger move when one or the other moved alone but would miss the left finger when both moved at the same time. If extinction were in scene-based reference frames, this pattern would be evident with observer rotation, as described above with neglect. The finger to the right in a rotated frame would be detected and the finger to the left would be extinguished, but only with bilateral movement conditions. Again, these patients often have trouble keeping their eyes fixated and must be reminded not to look in the direction they see movement and especially not to look toward their good side. Nevertheless, their eyes often tend to move in the direction reported, just as seen in patients with neglect. When fixation fails, their eyes typically move to the finger to the right of them when both fingers move simultaneously but to the left when only the finger to the left of the patient moves. This pattern of eye movements is also evident in scene-based frames. Clinical observations such as these demonstrate that there is little problem in attracting attention either to the left or the right within upright or rotated frames when only unilateral stimulation is present. They further demonstrate that when eye movements occur within reference frames, they follow a pattern consistent with the attentional deficit.

The discussion in this section to this point has been based on clinical observation, but there is ample experimental evidence in the cognitive neuropsychology literature that patients with neglect can utilize different frames of reference. To the extent that left neglect is due to a deficit in attending to the left side, this literature provides additional support that attention is guided by spatial reference frames defined by orientation and origin as calculated by the visual system.

In a relatively early study, Calvanio, Petrone, and Levine (1987) tested 10 patients with left neglect in an experiment that presented words in one of four quadrants on a display screen (4 trials in each quadrant) with the patient either sitting upright (aligned with the orientation of the words) or lying on their left or right side (90° clockwise or 90° counterclockwise from upright; Figure 3.14). Since the words remained upright in the environment, environmental and viewer-centered frames were dissociated. The patients were asked to read all the words they could. The mean number of words read are presented in Figure 3.14. Since there was a maximum of 4 trials presented in each quadrant, a perfect score would be 4. Although not all patients were perfect in reporting the words on the right side in the upright condition, the difference between right and left sides was clearly observed as shown in the upright condition of Figure 3.14. But the important question was what would happen in the two rotated conditions. I've placed the letter combinations of R and r and L and l in each quadrant of Figure 3.14, the first upper case letter designating left or right in environmental quadrants (the orientation defined by the letters on the page) and the second in lower case designating left or right in viewer quadrants (e.g., Rl refers to the right side defined by the display and the left side of the viewer). Of course in the upright display, environment and viewer left/right were coincident.

First notice the in the Rr quadrants patients were quite good in all head orientations and in the Ll quadrants they were poor. But what is most revealing is the consistency in the Rl and Lr conditions whether the patients were tilted right or left. In these conditions the number of words read were about the same (ranging from 2.1 to 2.6) and were in between

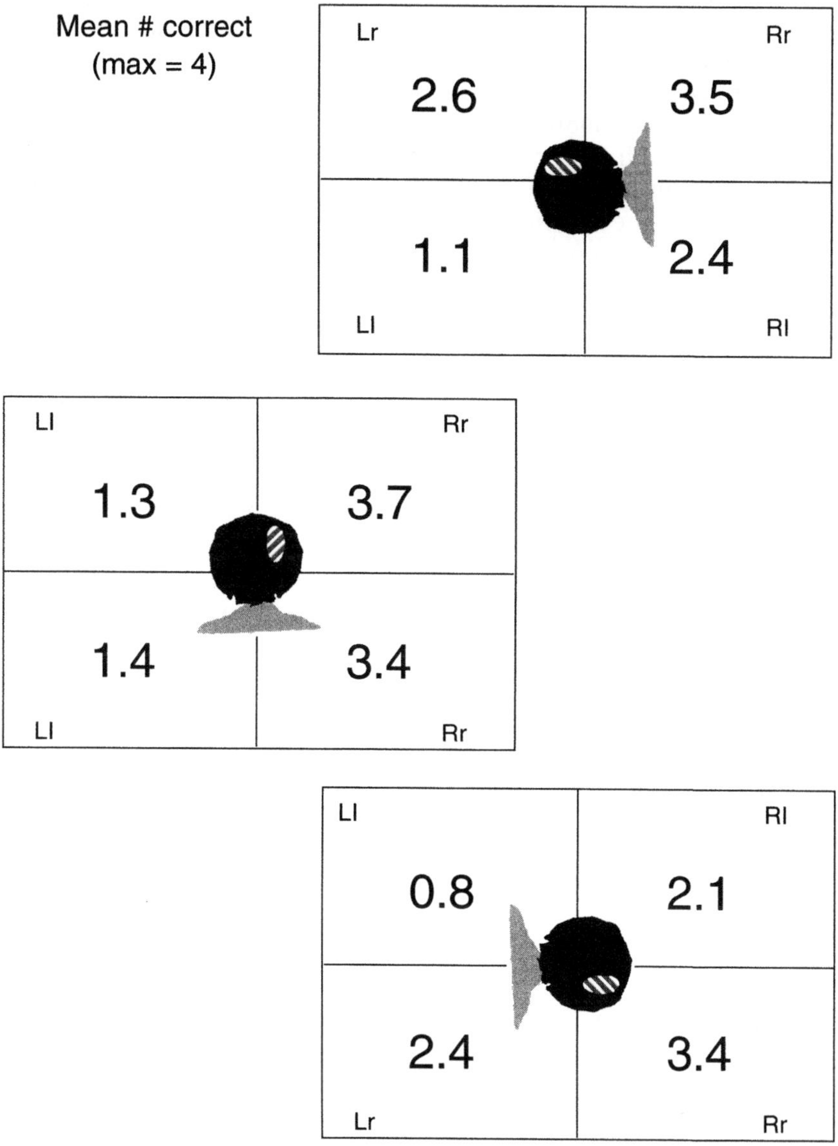

FIGURE 3.14. Mean number of words detected by a group of patients with left neglect when the patients were upright (represented in the middle of the figure) and when they were tilted 90° to the left or the right (represented on the top and bottom of the figure). The maximum number correct was 4. The uppercase R or L represents right or left in environment-centered coordinates, and the lowercase r or l represents right or left in viewer-centered coordinates.

the Ll (mean = .9) and Rr conditions (mean = 3.5). The combination of viewer and environment quadrants produced performance that was almost exactly in between the two extremes. Head and environment neglect were additive. These data show that both viewer and environment frames contributed about equally to neglect. The findings cannot resolve whether the two frames competed for attention on each trial or whether one frame dominated on one trial and another on another trial, but whichever is the case, the findings clearly indicated that neglect was not limited to viewer-centered coordinates and both frames influenced the pattern of results.

Other findings supporting the role of reference frames in attention deficits were reported at about the same time as Calvanio et al.'s study (1987) by Ladavas (1987). Ladavas (1987) tested patients with left extinction and demonstrated that targets that appeared in the left box of a cued pair of boxes arranged horizontally on a screen were detected more slowly and missed more often even when the targets were closer to fixation. For instance, when the box at location F in Figure 3.15 flashed to cue the subject that a target would likely appear there, a target appearing in an invalid location (E or G) was detected faster at G than at E, even though location E was closer to fixation (D). These effects could not be attributed to eye movements or eccentricity. Ladavas monitored all patients' eyes on every trial and eliminated trials on which they occurred.

Like the individual discussed in chapter 2 with mirror-image spatial performance studied by McCloskey and Rapp (2000), the origin of a spatial frame defined by the location of attention predicted the pattern of results observed in patients with extinction. Under most conditions, the locus of attention and that of fixation are the same, making it difficult to

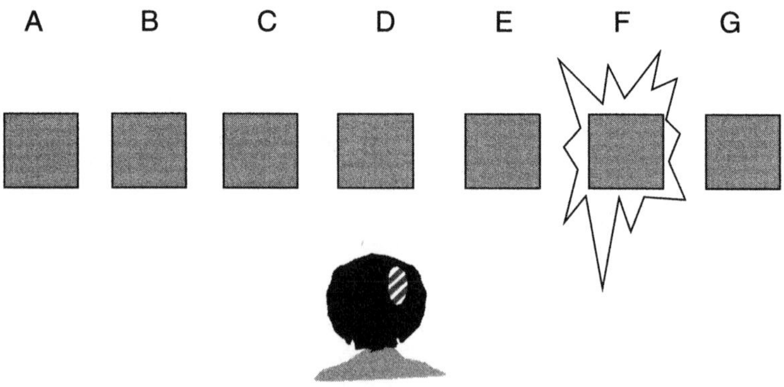

FIGURE 3.15. Positions of potential targets. See text for details. (Adapted from Ladavas, 1987.)

determine when effects can be attributed to retinal, viewer, or scene- and object-based spatial representations. But in the laboratory, the influence of spatial frames other than those defined by the viewer or the retina have been experimentally dissociated both by origin shifts and by rotation that dissociates frame orientations. These studies convincingly demonstrate that attention operates within a selected spatial frame of reference. Furthermore, they support other results discussed earlier showing that either attention or eye fixation can define the origin.

Many others have also documented neglect or extinction within spatial frames other than the viewer (e.g., Behrmann & Moscovitch, 1994; Behrmann & Tipper, 1999; Chatterjee, 1994; Driver, Baylis, Goodrich, & Rafal, 1994; Farah, Brunn, Wong, Wallace, & Carpenter, 1990; Karnath, Christ, & Hartje, 1993; Marshall & Halligan, 1989; Tipper & Behrmann, 1996), adding support for the earlier findings. Effects in extra-retinal, nonviewer-centered frames are often classified as "object-based." Although studies of this sort have clearly shown that attentional deficits can occur in different frames of reference, it is not always clear what investigators mean by object-based other than a frame that is not retinal or not viewer-based. Spatial attention within nonretinal frames has consistently been reported, but it appears to operate according to the same principles.

One debate about the underlying deficit in neglect and extinction concerns whether it reflects direct damage to part of an attentional system that distributes attention over space or affects the spatial frame itself. The distinction is one in which, for instance, left neglect due to right hemisphere involvement would reflect an alteration on the left side of a reference frame per se (over which attention is normally distributed) or a deficit in allocating attention to one side of an intact spatial frame. Theoretically, spatial attention could be intact but not able to move left because the space that supports attention is damaged or the spatial frame could be intact but attention could be "stuck" on the right. In fact, it may be the case that some cases of neglect affect the spatial representation, other cases affect attention, and still others affect both.

Edoardo Bisiach and his colleagues reported some of the earliest and best evidence for an underlying deficit in the spatial frame itself. In a well-known study, he showed that patients with left neglect missed landmarks on the left side of an Italian piazza relative to the perspective from which they imaged themselves looking at the piazza (Bisiach & Luzzatti, 1978). These authors argued that the space to support the left side was missing in their patients.

Another study from the same laboratory that is not referenced as often may be even more convincing in its support for directly altered spatial representations. Bisiach, Luzzatti, and Perani (1979) placed one cloud-

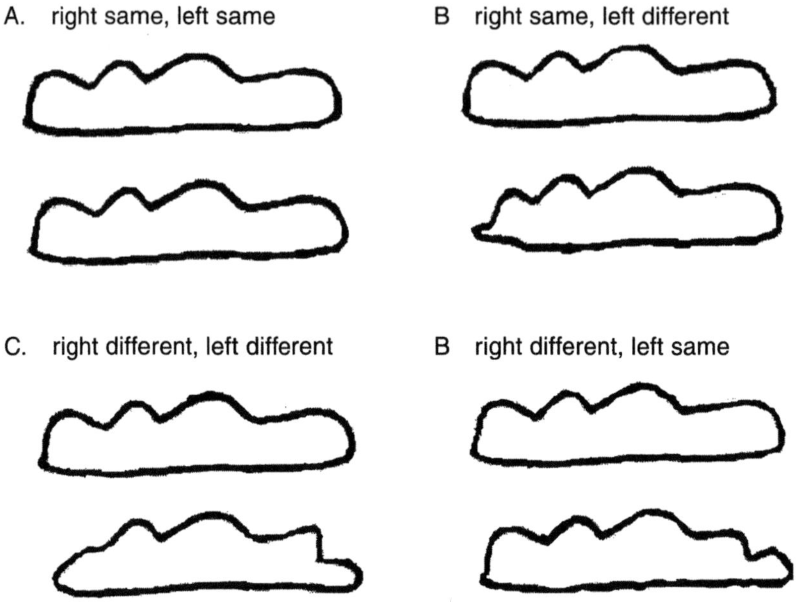

FIGURE 3.16. Examples of cloud-like stimulus pairs that are either the same on both sides (a), different on the left (neglected) side but the same on the right (b), different on both sides (c), or different on the right but the same on the left (d). (Adapted from Bisiach et al., 1979.)

like shape above another cloud-like shape and asked patients to report whether the two clouds were the same or different (Figure 3.16). When the two clouds were the same on the right side, the patients reported that they were the same whether or not they were the same on the neglected left side (a and b). Likewise, when the two clouds were different on the right, the patients reported that they were different whether or not they were the same on the neglected left (c and d).

All patients had left neglect, so this finding was not surprising, but what came next was. Bisiach placed a flat barrier with a central slit in it between the clouds and the patients and then drifted a cloud pair rightward or leftward behind the barrier. At any given time, all a patient saw was the parts of the pair showing through the slit (Figure 3.17). Even though the patients were not exposed to the cloud pairs in full view, they performed the same as before. They reported the clouds as same or different when they were the same or different on the right side irrespective of whether or not they were the same on the neglected left side. In order for this to happen, the representation of the clouds must have been reconstructed by the patients as the stimulus pair passed behind the slit. What was missing was a code for the left side of the resulting mental representation.

FIGURE 3.17. Example of what the patients in the Bisiach et al. (1979) study would have seen at a given moment as the cloud-like pairs shown in Figure 3.16 were drifted behind a slit in a barrier.

This procedure revealed that the left side of the stimulus pair was neglected just as if it had been presented in full view despite the fact that the left side of the figures was presented in the same place in the stimulus as the right side (right or left drift also made no difference). The data demonstrated that the left side of a stimulus pair that was never shown on the left side of the display or to the left of the viewer still could be neglected. The slit in the barrier was in the center where the patients were looking (i.e., attending), aligned with a viewer-centered, gravity-centered, object-centered, and barrier- (or scene-) centered frame. The spatial representation of the clouds was best accounted for by an internally generated spatial reference frame that could not represent the space of the left side of the cloud pairs. There was essentially no place to hang parts on the left side even though the features on the left were clearly perceptually processed during initial viewing.

Several later studies demonstrated that stimuli presented in the neglected space can be implicitly encoded and affect performance in attended space, but whether the *locations* of the stimuli that affect performance are encoded in the correct locations is not known. I will return to this issue in a later chapter when I talk about implicit and explicit space and object representations, but for the present purposes, these findings very strongly favor the necessity for some type of mental spatial representation to account for the results reported by Bisiach and his colleagues.

Findings such as the ones I have discussed in this section show that whatever representation is selected for further processing, spatial information that supports that representation is also required to bring the information to awareness. This conclusion should not be construed as claiming that all cases of unilateral neglect are due to the direct loss of part of a spatial reference frame. Neglect comes in many forms, and some cases may be due to direct damage to spatial representations, while others may reflect the direct loss of spatial attentional processes. This issue will be revisited in chapter 7.

☐ Spatial Extent, Spatial Resolution, and Attention

Spatial location refers to anything from the position of a finite point to a region of greater or lesser size. Our solar system has a location within the universe. From the perspective of the universe, the solar system is a small dot of little importance, but from our viewpoint it is rather large. In fact, it is difficult to think of the solar system as having a location at all except when we imagine it in the broader context or universal frame of reference (e.g., our galaxy or the universe as a whole). So when we speak of attending to a location, the next question might be how much space do we mean and relative to what. Locations within spatial frames can, of course, be defined as points (e.g., the origin, the intersection of $x+1$ and $y+1$), or they can be defined as a large region of a particular size and shape (what might even be called an object). Alternatively, they can be defined by a type of Gaussian distribution where attentional resources are distributed with a defined point being the peak and falling off gradually around this peak.

The area of space over which attention is distributed is often referred to as the "window of attention" and is sometimes likened to a spotlight where the beam magnifies the center of the window with the borders fading off gradually. These metaphors have had a significant influence on studies of spatial attention in cognitive science as well as cognitive neuroscience. It is common to use such terms whether describing functional imaging activity (fMRI, PET) or purely behavioral data.

In fact, how attention selects a spatial region is a well-studied question in the literature on spatial attention. But again, the space that is selected has been assumed to be the one space we typically think of as out there. However, whether speaking of a single space of the space within any given frame of reference, the issue remains of how the parameters of the attentional window are determined. Can spatial attention be directed to a single point, and if not how small is the region it can attain (e.g., Eriksen & Yeh, 1985)? Is spatial attention best modeled as a gradient (Jonides,

1993; LaBerge & Brown, 1989) or a spotlight (Posner, Snyder, & Davidson, 1980), or is it more like the aperture of a camera that zooms in and out (e.g., Eriksen & St. James, 1986)?

There is ample evidence that spatial attention can be constricted to a small area of space or distributed over a larger area. Its distribution can be changed by bottom-up information such as the organization of objects by such things as grouping or by top-down control, as occurs when inhibiting irrelevant items that flank a target (Eriksen & Eriksen, 1974). Its shape and distribution can be affected by the task as well, such as that observed during reading (Rayner, McConkie, & Zola, 1980). There seems to be a flexible size and form over which spatial attention can enhance information processing.

Some years ago LaBerge (1990) proposed an elegant neurobiological model for controlling the size of the attentional window that relied on signals from the pulvinar of the thalamus (a very old structure) interacting with the parietal lobes. The model was partially based on functional imaging data demonstrating increased activity in the thalamus when the task required attention to be narrowed to a central part of a stimulus versus when no adjustments were necessary to perform the task. Given the evidence for parietal function in attending to locations in the visual field, the addition of thalamic modulation offered a neurobiological theory of how the size of the window around a cued location could be determined.

There is also convincing evidence from electrophysiological data recorded from the temporal cortex of monkeys that neural responses in areas of the temproal lobe can be modulated in a way that appears to expand and contract attended regions of space. The now classical work by Robert Desimone and his colleagues has shown that the cellular firing rate over an area of space can change the response profile of a neuron depending on attentional manipulations (Moran & Desimone, 1985). They recorded from single neurons in monkey cortex (V4) and demonstrated that the receptive field size (i.e., the area over the visual field to which a neuron responds to a preferred stimulus above some preset threshold) could essentially "shrink" when a to-be-ignored distractor was placed within its field along with a to-be-attended target. A stimulus of a given type could change the pattern of spike frequency over baseline, essentially enlarging or constricting the spatial window of a single cell (i.e., its receptive field size). However, in terms of functional anatomy, the question is where the signal that modulates receptive field size is generated. A cell cannot tell itself to change the area over which it fires. The source of the modulation must come from outside the cell. A potential source is from the dorsal spatial pathway of the cortex that includes both frontal and parietal areas, the "where" processing stream (Desimone, 2000; Mishin, Ungerleider, & Macko, 1983).

In fact, more recent findings from Desimone's laboratory have shown that filtering out distractors is decreased by lesions in the temporal lobe of monkeys in areas V4 and in more anterior sites in the temporal lobe known as TE (DeWeerd, Peralta, Desimone, & Ungerleider, 1999). These findings have been confirmed in humans by testing a patient with a lesion in V4 using the same paradigm as with monkeys (Gallant, Shoup, & Mazer, 2000). When distractor contrast increased, making the distractors more salient, the ability to discriminate targets suffered with lesions in these temporal areas. More recently, Friedman-Hill, Robertson, Ungerleider, and Desimone (2003) demonstrated that parietal lesions in humans affected filtering in the same way, again using the same methods. These results are consistent with interactions between dorsal and ventral visual areas that form a network in which the parietal lobes are part of the source (perhaps linked to the thalamus) of the signal that filters out distractors, and temporal areas are the receivers. For normal perceivers, distractor filtering changes the form and size of the spatial window of attention through these interactions. With damage to either the transmission source or the receiver, the effects will be the same, namely, deficits in setting the size of the spatial window and increasing the influence of distracting stimuli.

This brief overview gives the flavor of the convergence between the cognitive and neurobiological literature on issues of the size of a region over which attention is spread. However, there is more to spatial attention than selecting the size of a region in the visual field over which to allocate resources. This is the case whether talking about large areas that different hemispheres monitor (right visual field by the left hemisphere or left visual field by the right hemisphere) or small areas that single neurons monitor (their receptive field size).

Spatial Resolution

Besides the obvious 3-D spatial structure that must be resolved by the brain from a 2-D projection on the retina, there is also the resolution or grain that must be considered. For instance, some old movies appear as if sand had been ground into the film, making the grain appear course. The picture can look somewhat blurry and the details difficult to see. On the other hand, a new DVD version provides a crisp, clear picture due to the higher spatial resolution. "Due to" is not quite correct, because of course the seeing is not being done by the technology, but by the brain. The brain encodes a range of spatial resolution in a visual scene. Early sensory vision and primary cortex carry information about the spatial frequencies in the stimulus (as measured by the cycles per degree of visual angle) in a number of "spatial frequency channels" (DeValois & DeValois, 1988). The grainy look of an old movie occurs because high spatial frequency channels are not stimulated (because the information is not there to activate them) and thus provide no information for the visual system to resolve or attend to finer spatial scale. However, lower spatial frequency channels are stimulated, and the resulting percept is of a somewhat blurry, rough-grained picture. In a DVD picture both higher and lower frequency channels are activated, providing spatial information across a wide range of spatial resolution that results in a clearer picture.

The computations that utilize spatial frequency per se happen preattentively (before attention), yet we can choose to focus on the courser or finer grain of a stimulus (Graham, Kramer, & Haber, 1985). In terms of properties of stimuli we see, we can pay attention to the texture of an object or to its overall form (see Kimchi & Palmer, 1982). There is good evidence showing that attention can modulate spatial frequency detection (Braun, Koch, Lee, & Itti, 2001; Davis & Graham, 1980; Yeshurun & Carrasco, 1999). Attentional selection of some frequency channels is not limited to vision. There is also good evidence for similar channel selection for auditory frequency (Johnson & Hafter, 1980). One mechanism that we call attention modulates another that we call a channel.

The result of this engineering is that sensory information is encoded at multiple spatial resolutions, with attention choosing the ones that are most appropriate at the moment. Similarly, information in neural channels is present across the spatial spectrum, and attention can selectively attend to the channels that carry the signal that is most useful for the task. One could metaphorically relate this system to something like a ruler, where attention may focus on feet or inches. When attending to a 1 foot patch, the ruler as a whole becomes the object of attention (i.e., attending to lower spatial resolution), but when attending to 12 inches, inches become the object of attention and higher resolution is necessary. Both scales are always present in the ruler (i.e., spatially represented by a reference frame), but information is amplified or dampened depending on how useful a particular unit size is for the task. This architecture also allows fast switching between one level of spatial resolution and another and has been invoked to account for changes in the time to perceive global and local properties of a stimulus (Ivry & Robertson, 1998).

As one can see, spatial attention is involved in determining both the area over which attention will be allocated and the spatial resolution needed. Although these two properties of spatial scale can affect each other (e.g., a smaller attentional window favors higher spatial frequency), there is evidence that they are represented separately. For instance, visual aftereffects that appear after viewing gratings of black and white stripes change the perceived width of each of the stripes but do not change the

FIGURE 3.18. Does one pair of faces seems slightly larger than the other?

perceived overall area that the stripes cover (Blakemore & Sutton, 1969). After adapting to a grating with thin stripes, the stripes in another grating are perceived as slightly thicker, but the region in which the gratings are shown does not expand or contract. On the other hand, the spatial frequency content of a stimulus can be the same, but the perceived size may change. For instance, the faces in Figure 3.18 are the same in terms of spatial frequency spectrum (only changing in contrast), but the white faces on a dark background are usually perceived as slightly larger than the dark faces on a white background.

□ Spatial Resolution and Reference Frames

It is easy to find examples of spatially constricting and expanding attention in a selected spatial reference frame. A narrow window is better when proofreading this page than when counting the number of words or lines. Adjustments in spatial resolution are also helpful. When proofreading, attention to higher spatial frequencies would be more beneficial than attention to low.

Spatial resolution may also influence frame selection itself. If an elephant appeared in peripheral vision, the frame of this page might be rendered relatively unimportant, and the selection of a new frame that is more panoramic would seem reasonable. Switching from the more local frame of this page to the more global frame of the environment seems like a good strategy under these circumstances. Given the visual system's lower spatial resolution in the periphery, could it be that switching frames under these circumstances corresponds to switching between spatial frequency channels? In fact, there is good evidence that spatial frequency may contribute to frame selection within the hierarchy of spatial frames available in normal visual environments.

Spatial Resolution and Global/Local Frames of Reference

One way to examine the role of different features in frame selection is to examine how switching between frames is influenced by manipulations that affect these features. Repetition priming methods used in several experiments have demonstrated that there is a cost associated with switching from one frame to another (see below), just as there is a cost from switching from one location to another within any given frame (Posner, 1980). This switching cost can be ameliorated by variations in spatial frequency or spatial resolution of the stimuli.

Repetition priming is a method often used to determine the type of representation that persists to influence later performance. Its use is ubiquitous in the cognitive literature, and it is a powerful method that has often been used to study various attentional and memory components. Part of its power is that it allows for inferences about what representations were created and/or what processing occurred at the time the previous stimulus (prime) was presented. Responses to the second stimulus (probe) indirectly reveal what these might be.

More often than not, the emphasis has been on the nature of the representation that persists over time. If a stimulus is stored adequately in memory, it will improve performance if that stimulus or one similar to it is repeated (Scarborough, Gerard, & Cortese, 1977). If a shape is represented as a spatially invariant object, performance will be better when the shape is presented again, even if it changes location and/or reflection (Biederman & Cooper, 1991; but see also Robertson, 1995). If attention selects one of two shapes in a stimulus on one trial, the selected shape will improve performance when it is repeated and the unselected shape will worsen performance when it is repeated (Figure 3.19). The worsening of performance is known as "negative priming" (Allport, Tipper, & Chmiel, 1985) and is believed to represent inhibition of the ignored shape leading to worse performance later on (Figure 3.20).

Another way of thinking about repetition priming in studies of attention is in terms of attentional weights created from a previous act of at-

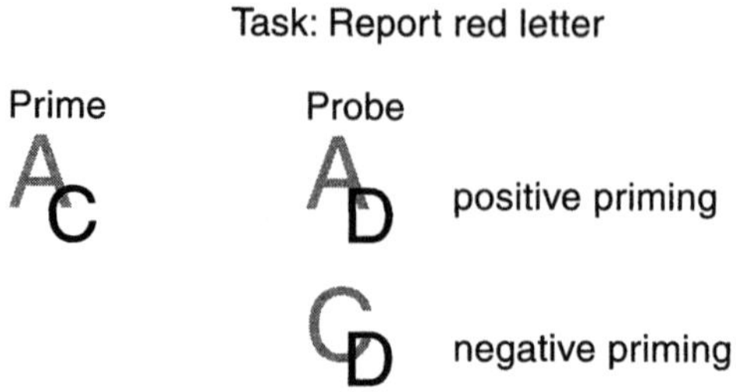

FIGURE 3.19. A typical negative priming paradigm might be to report the red (represented by gray) letter in a pair of two overlapping letters of different colors in a series of prime/probe trials. In the figure, the A is the target in the prime, and when it later appears as the target in the probe, performance is facilitated (positive priming), but when the distractor in the prime becomes the target in the probe, performance is worse (negative priming).

tending (Robertson, 1996; Wolfe, 1994). For instance, in negative priming, both shapes may be represented with equal strength but could be tagged as the "right" or "wrong" shape when processing the prime stimulus. When the wrong shape (the one that was inhibited before) then appears as the right shape (the one that now requires attention), the system must adjust to the new contingencies. This adjustment will take time and effort and lead to slower identification and/or errors. This hypothesis predicts that if the wrong shape continues to be the wrong shape on the probe trial (i.e., the one that required inhibition), then subjects will be better when it requires inhibition again. Allport, Tipper, and Chmeil (1985) and Neumann and DeSchepper (1992) found evidence that this was the case. When a target letter was paired with a nontarget letter, there was positive priming when the same letter appeared as a target in a subsequent trial, and there was also positive priming when the distractor letter in the prime appeared as the same distractor letter in the probe. The act of inhibiting the distractor on the first trial enhanced the ability to inhibit it again on the subsequent trial. It was the attentional process that operated on the letters (whether target or distractor) that improved performance, not the strength of letter representation per se (see Salo, Robertson, & Nordahl, 1996 for a similar finding and interpretation using the Stroop task).

This type of approach can also be applied to some findings about spatial attention. Selectively attending to a target in one location on one trial speeds selection of a target in the same location on the next trial, and distractors that are presented in the same location also increase selection speed. The processes of both facilitation and inhibition are sustained over time. In addition, both effects are cumulative over trials (Maljkovik & Nakayama, 1994).

Perhaps somewhat more relevant for the topic of reference frame selection is a set of experiments with global and local levels. Studies have repeatedly shown that selecting a target at one level (either global or local) facilitates selection at the same level on the next trial but slows selection when the target changes levels (Robertson, 1996; L. M. Ward, 1982). Even more importantly, this effect is independent of whether the target shapes themselves change or not (e.g., if both E and S are targets, it does not matter if the shape is repeated; rather, it matters whether the target is at the attended level).

Since the level-priming effects are relevant to issues concerning selection of spatial reference frames that are more global or more local, a bit more detail seems in order. In the key experiment, subjects were presented with a hierarchically constructed stimulus (see Figure 3.20) and were told to press one key with one hand if an H appeared and another key with the other hand if an S appeared. On each trial there was always an H or an S and it could appear either at the global or local level but never at both. Unbeknownst to the subjects, the trials were arranged into prime–probe pairs so that there were an equal number of trials where the target level remained the same and when it changed. When the target was at the same level, response times were faster than when it changed, and this occurred whether the target letter itself (and thus the response) changed or not. Also, the effects were symmetrical. The difference between same level and changed level target detection was the same whether the change was to the local from global level or to the global from local level. This symmetry has been replicated several times (N. Kim, Ivry, & Robertson, 1999; Lamb, Yund, & Pond 1999; Filioteo, Friedrich, & Striker, 2001; Robertson, Egly, Lamb, & Kerth, 1993: L. R. Ward, 1982). Further studies have shown that these priming effects are related to the different spatial frequencies that can be used to parse levels (Robertson, 1996; 1999; although see Lamb, Yund, & Pond, 1999), are not location specific, and last at least 3 seconds without any reduction in strength.

Attentional Prints

Basically, when the act of selection successfully revealed a target at one level (whether global or local), that level received more attentional weight and facilitated the next act of selection at that level. There was the formation of what I have called an "attentional print" that marked the spatial scales that had been attended on a previous trial.

Although I have talked about these results in spatial resolution terms, the global and local level of a hierarchical stimulus like that in Figure 3.20 can be thought of as two objects (shapes) or two spatial frames in

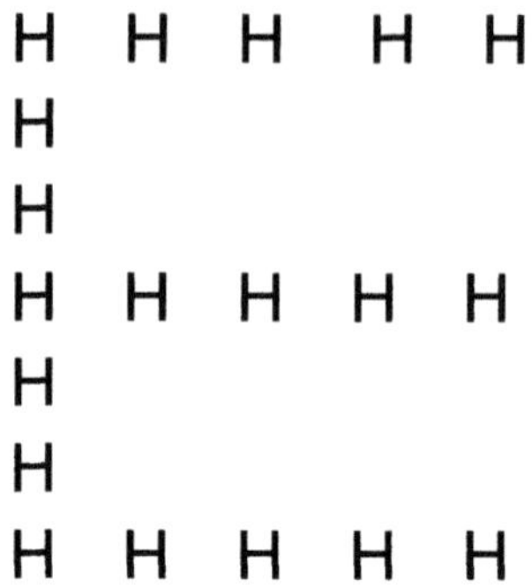

FIGURE 3.20. Example of a hierarchical stimulus with a global E created from local Hs. In the example in the text, H would be the target.

any one stimulus presentation. By using repetition priming methods, I was able to determine that it was the spatial resolution that determined priming in this case. The level-priming effect occurred whether the target remained the same or changed from trial to trial.

A mechanism that supports something like an attentional print would seem highly beneficial in everyday life. When reading the words on a page we want to stay in the same frame with about the same visual resolution as we move attention from one word to the next. When watching a football game, a more global frame may be desirable in order to appreciate the plays. Every time we look away from and back to the game we should not have to reconstruct the spatial organization of the field and the players. Instead, there is a sustained code that tags the best spatial resolution for that stimulus according to the control settings from the previous act of attending.

Other features of spatial coding appear to retain a similar trace. For instance, McCarley and He (2001) used stereopsis to vary the orientation of 3-D spatial planes in depth and then asked subjects to detect a target in the central plane of the display when it appeared as oriented toward the ceiling or the ground (see Figure 3.21). Priming effects were analyzed to determine whether search time was affected by the orientation of the plane or by the display as it was projected onto the retina. Search was facilitated within a plane (i.e., spatial frame defined along a 3-D projection). More importantly for the present discussion, when sequential trials were both ceiling-like or both ground-like search was faster than when the stimulus as a whole changed from one to the other. Although the origin and unit size of the selected plane remained the same, perceived orientation varied, creating the need to change the frame in which search proceeded.

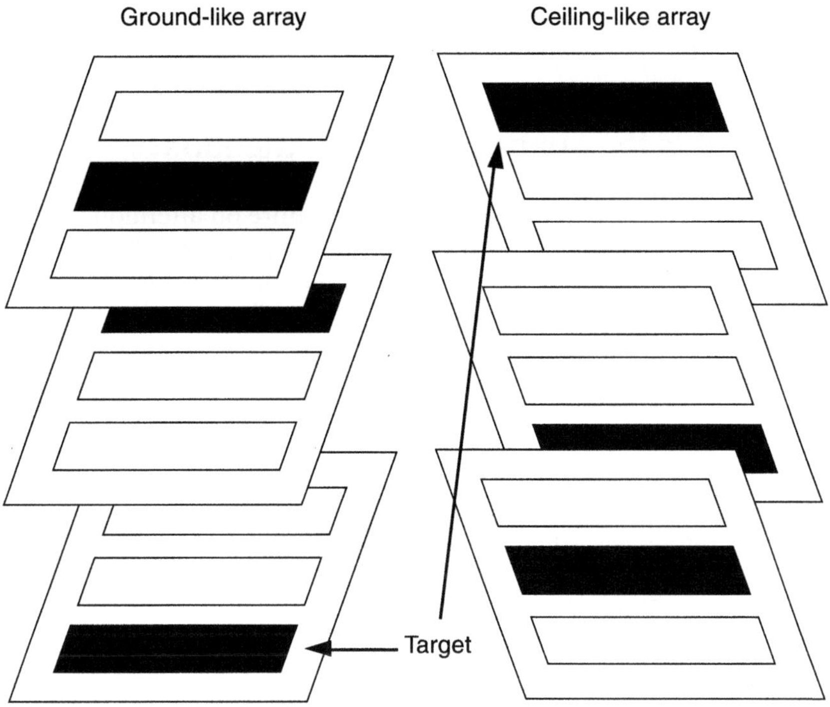

FIGURE 3.31. Example of the types of planes the subjects would see. Target detection was better when the planes were perceived as separated in depth, as shown. (Adapted from McCarley & He, 2001.)

Another study by Sanocki and Epstein (2000) directly tested the question of whether a spatial frame alone could prime subsequent judgments of items that did not appear in the priming scene, and indeed it could. Even an impoverished sketch that gave the spatial layout of a scene produced positive priming for items that were not in the sketch as long as it provided adequate information to construct a spatial framework.

These studies were not designed to test the relationship between spatial scale and reference frames directly, but they do support the value of spatial frames in guiding attention and the importance of frame selection in determining the ease in finding a desired object in a cluttered array. Priming within different levels of hierarchical shapes and different depth planes seems to rely, at least in part, on the spatial resolution as well as other spatial properties of selected frames.

Attention does more than simply move around the space we perceive. It is involved in frame selection, selection of spatial resolution, establishing the window of attention over the reference frame it is operating within and keeping a trace of the selection process and the features and frames that resulted in a previous act of selection.

What Is the Space for Spatial Attention?

Often when I listen to a talk or read the literature on attention, I get the impression that most investigators agree on what space is. This seems to be the case whether they study the distribution of spatial attention or whether they describe the effects of spatial attention on other processing mechanisms such as those involved in object perception, visual search, or even eye movements. Although there are debates (sometimes raging) within the visual sciences about how space is computed (e.g., by Fourier analysis of spatially tuned frequency channels, lines and angles, overlapping receptive fields, etc.), these debates are generally limited to the representation of space itself and not to how attention might contribute to and select the spatial structure that emerges. The assumption seems to be that attention can be agnostic to whatever it is that allows for the computation of perceptual space itself. A unified spatial map of the world is generated (the one that we know), then spatial attention simply uses that map.

I am overstating the case, but in fact most investigations of spatial attention do not define what space means in any given context, and it appears to mean different things in different papers. For some, space is measured in retinal coordinates. Receptive fields of single visual neurons is one example. A receptive field size is by definition the size of an area measured to which a neuron fires above some baseline. Attention has been said to modulate receptive field size (Moran & Desimone, 1985), although this way of speaking is somewhat loose. When a monkey attends to a stimulus with a target and distractor in the receptive field of the recorded cell, a location within the cell where the distractor had previously increased firing rate when presented alone might now show baseline firing or even decreased firing. It is as if the window of attention for that neuron had shrunk.

Clearly vision must begin at the retina, but it is also clear from the many examples I've discussed throughout this book that it soon goes beyond retinal parameters. Defining the space for spatial attention in terms of retinal space (as is often done implicitly) is not sufficient. Eye movements, body rotations, and visual motion all change retinal location, and it seems that any animal would be better off if attention used less easily disrupted spaces.

Investigators enslaved to retinal coordinates are not limited to many of those who study single units in animals but also include those who present stimuli in the right or left visual field to study hemisphere laterality in normal perceivers. In this case the space is the whole left or whole right side relative to a vertical line through fixation.

Another common assumption about space is that it conforms to the spatial structure of the world. In other words, if the distance between x and y is the same as the distance between x and z in the external world, this relationship is assumed to hold for the allocation of spatial attention (e.g., Figure 3.22). If it does not, then typically the conclusion is that attention is responding to something other than space (e.g., object-based attention). This leads to the idea that attention selects locations in one spatial map that represents space as we know it, and selects everything else in a map that represent stimulus features or a collection of elements, generally referred to as objects.

A notable exception to the space-as-unitary assumption is the egocentric/allocentric distinction derived from the neuropsychological literature (see Milner & Goodale, 1995). Egocentric refers to space within the action space of the body, and allocentric refers to space at more of a distance. These spaces are orthogonal to object/space hierarchies, as these

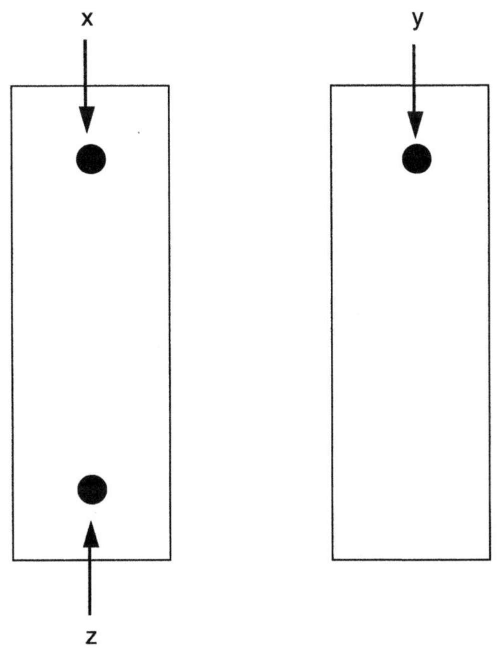

FIGURE 3.22. The distance between x and y and x and z are the same, but attention moves from x to z faster than from x to y. This violation of space as measured on the page is normally invoked as evidence for object-based attention. (Adapted from Egly, Driver, & Rafal, 1994.)

hierarchies can exist within both proximal and distal spaces. Nevertheless, this is one example where at least two types of spatial representations have been proposed based on two different uses (action and perception).

Others talk about spatial processing channels. As discussed previously, there is convincing psychophysical and neurobiological evidence for spatial frequency channels that process information at different spatial resolutions in early vision. The number of channels has been debated, but it is generally believed to be small, possibly as small as 3 (see Graham, 1981), but probably somewhat more. Some have argued that the spatial map that we visually experience is computed from the orientation and spatial frequency information carried in these channels. In this view space is a construct of luminance contrasts in frequency space. The strong conclusion is that spatial maps do not exist without luminance contrast information (i.e., without something in the visual field). However, even in a Ganzfeld field attention can be directed to, say, a location in the upper left quadrant, just as it can be directed to a location within the homogeneous clear blue sky. Is this what is meant by spatial attention? Does it only exist in its pure form when no contrast edges are present in the scene? When one takes the logic to the extreme, the question of what is the space for spatial attention only applies to a Ganzfeld field, but as the discussions throughout this chapter make clear, this cannot be right.

In sum, a great deal of work in cognitive psychology, cognitive science, and neuropsychology and neurobiology over the past few decades has uncovered a number of principals regarding spatial attention. Components of spatial attention have been isolated through well-controlled studies, and we know a great deal about the ways in which attention is distributed over the space that we see when searching for the objects we seek. We also know something about the neurobiological mechanisms that are necessary for normal attentional performance. Along the way we have discovered interesting and important facts about patients with spatial attentional problems that have had an impact on understanding these deficits, and this in turn has led to new diagnostic and rehabilitation efforts. Overall, this area of study reflects great success.

Nevertheless, it is not at all clear that everyone who studies spatial attention is talking about the same space. There is growing evidence that there are multiple spatial maps in which attention can be distributed, and the selection of these maps themselves appears to require an attentional act. It is not sufficient to think of spatial attention as tied to the retina or the viewer on the one hand and to the external world on the other. Nor is it sufficient to call anything other than viewer- or retinally defined space object-based. This issue of objects and object-based attention will be explored in the next chapter.

4

CHAPTER

Object-Based Attention and Spatial Maps

Objects in the environment exist in different locations. In turn, parts of objects take their place at different locations within an object, and parts themselves have spatial structure. A simple rule of nature is that no two objects can exist in the same location at the same time, and if they attempt to do so, there will be a rather substantial reconfiguration. Since the visual system evolved in this world and not in some other, it would be surprising if our perception of space and objects did not somehow reflect these natural principles. Even when overlapping figures are presented on the same plane, as in Figure 4.1, the visual system parses them into perceptual units in different spatial planes so that one unit is either perceived as in front of or behind the other. They are not in the same space in our mind's eye even when they are in the same space on the page. The rules of perception are such that the perceptual world is isomorphic to the physical environment only as closely as is sufficient to support survival.

This isomorphism between the structure of object/space in the external world and the internal representation of that world makes it very difficult to design experiments to determine when or even whether attention selects spatial locations or objects, a fundamental question in the attention literature today (see Yantis & Serences, 2003; Vecara & Farah, 1994). Early attempts to sort out whether attention was allocated to spatial locations or to the objects that inhabited them supported object-based selection

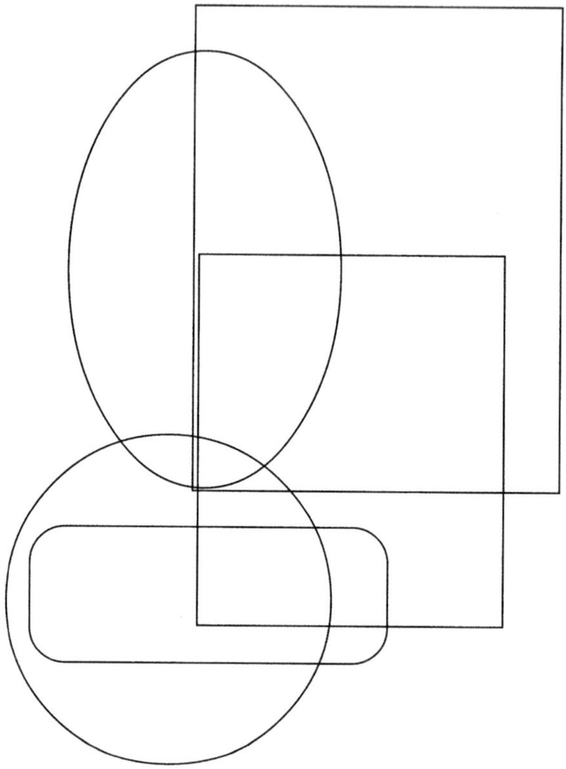

FIGURE 4.1. The visual system adds depth to this figure, resulting in the perception of a selected shape as figure in a different plane.

(Duncan; 1984; Rock & Guttman, 1981). Several studies demonstrated that reporting two features from the same object was faster than reporting two features from different objects. Nevertheless, because objects in these studies inhabited different spaces, it was difficult to know whether attention had selected the object or the spatial area it covered. A feature from a different object was in a different location. Recent studies have attempted to overcome this problem by presenting stimuli in the same spatial relationship to each other and either rotating the stimuli out of alignment with a cued location or measuring how attention moves within an object versus between two objects when the distances are equated (e.g., Egly, Driver, & Rafal, 1994; Kramer & Watson, 1996; Ro & Rafal, 1999; Tipper, Weaver, Jerreat, & Burak, 1994). These studies have generally obtained both space-based and object-based attentional effects, leading to the general consensus that there are both space-based and object-based attentional mechanisms.

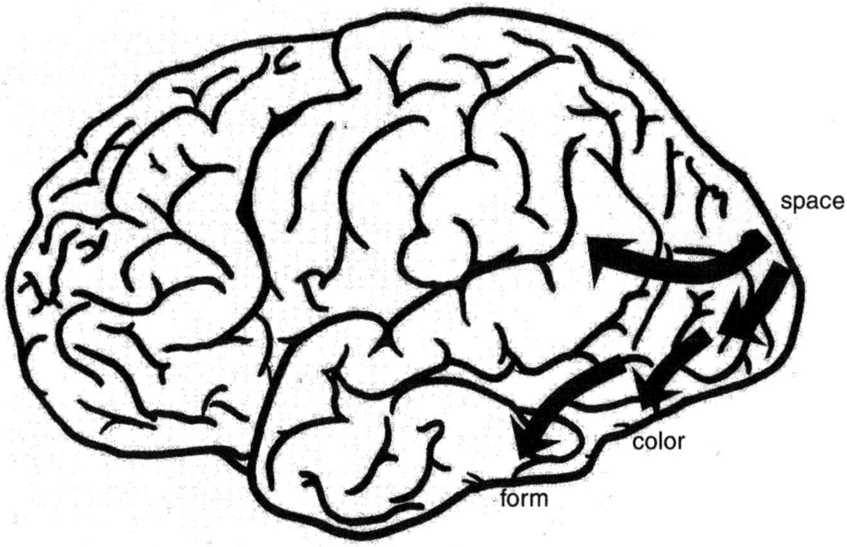

FIGURE 4.2. A dorsal processing stream is thought to process space to determine "where" or "how," and a ventral processing stream is thought to process features to determine "what."

This idea has been augmented by neurobiological evidence for two separate processing streams in the cortex (Figure 4.2): a dorsal system involved in space processing and a ventral one involved in processing objects and their constituent features (Ungerleider & Mishkin, 1982). The fact that damage to dorsal areas (especially parietal lobes) produces spatial deficits while damage to ventral areas produces visual agnosias (i.e., object recognition deficits) adds substantial support for the object- versus space-based distinction (see Farah, 1990).

There is no doubt that dorsal and ventral streams process different information, but the conclusion that objects are selected by one stream independent of their locations while locations are selected by another independent of objects is not as logically consistent as one might like. Objects have a spatial structure, and again, natural scenes contain hierarchically organized objects, with each level in the hierarchy defined by its own space. There are multiple levels of object/spaces that the visual system deals with successfully on a full-time basis.

We are all familiar with the experience of seeing where something is even when we do not know what it is (although what we see might be just a smudge of some sort that, if asked, we would report as a smudge), but few of us have experienced seeing what something is without knowing where it is. Nevertheless, this does happen when lesions are located in specific areas. As described in chapter 1, seeing an object but not its spatial location is what the world looks like to a patient with Balint's syndrome. This syndrome is produced by bilateral parietal lesions or damage that affects functioning in both parietal lobes (lesions in the dorsal cortical stream of processing). These patients perceive a single object (it might be small or large, complex or simple at any given time), yet they have no idea where it is located. It is not mislocated. Instead it seems to have no position at all. Attending to the object appears to be intact but attending to its spatial location is not.

Cases like these are very compelling in their surface support for object- versus space-based attention, but there is a problem. How can a person without a spatial representation of the external world perceive even one object when objects are defined by their own spatial structures? A face is not a face unless the features are in their proper locations relative to each other, yet a person with Balint's syndrome has no difficulty in recognizing faces. A table has a top attached perpendicular to legs that support it. How can a person who loses space see a table without perceiving the spatial relationships between of its parts?

The most prevalent theories of object- and space-based attention rely on the idea that perception works out what is to be considered an "object," and then attention selects either the object or its spatial location. A few researchers have gone one step further to suggest that the objects define a set of hierarchically arranged representations, and attention is used to select the object in this hierarchy (see Baylis & Driver, 1993; Watt, 1988, for early starts on this idea). But evidence discussed in chapter 3 (Rhodes & Robertson, 2002; Robertson 1996) demonstrate that spatial reference frames can be selected and set in place before objects are even presented and thus before objects are selected. The selected reference frame then guides the distribution of attention. In other words, attention does not necessarily select after the world has already been parsed and analyzed by object-based systems. Rather, object-based and space-based systems seem to interact at a very early stage. Nevertheless, there is a large body of evidence leading to claims that attention is object-based, and some of the major support for these claims will be the topic of the next sections.

□ Dissociating Object- and Space-Based Attention

One of the methods that has been used to overcome the challenge posed by the fact that objects and their spatial locations are integrally linked is to use motion to move objects from a cued location and then determine whether attention moves with the object or remains at the cued position. The prediction seems intuitively obvious. We track objects in the world, and it would be maladaptive to maintain attention at the location from which, say, a lion just moved when it is the lion that is meaningful. Nevertheless, when the lion moves, so does its relative location (e.g., to the observer, to the background in the environment, to other lions), so how can we tell whether it is the lion or the space the lion drags with it that is the object (so to speak) of attention?

Attentional Inhibition of Objects (IOR)

Several investigators have developed fairly clever ways to address this question. For instance, Steve Tipper and his colleagues used an exogenous Posner cuing paradigm (one in which the cue was nonpredictive and provided no motivation to control attentional allocation) followed by rotation (see Tipper & Weaver, 1998). A target was then presented either in the same location as cued or in the same object (Figure 4.3). Objects were defined as each of the individual squares. Given that stimulus

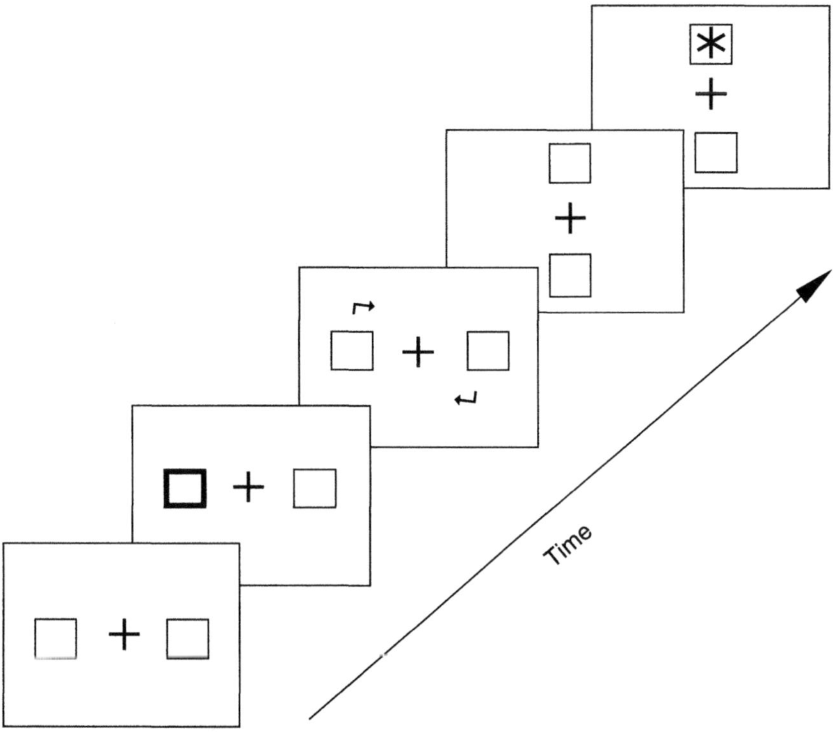

FIGURE 4.3. Example of a trial testing for object-based attention in a variation of the Posner cuing paradigm. In this example the target (*) appears in the cued object. (Adapted from Tipper et al., 1994.)

rotation occurred between cue and target to dissociate the cued location from the cued object, a relatively long delay between cue onset and target onset (stimulus onset asynchrony, or SOA) was necessary. Rotation appeared smooth and was 90° from the starting position.

Before going on to discuss the results, a few facts should be kept in mind. At longer SOAs, the normal benefit for cued locations changes to a cost, at least when nonpredictive cues are used (Figure 4.4), and even when there is no rotation (Posner, 1980). This pattern is believed to represent early attentional facilitation and later inhibition of the cued location. The later phase is often referred to as inhibition of return, or IOR, because it is thought to drive the movement of attention to objects or spatial locations that have not previously been attended and to reduce the probability of returning to an object or location that has been already attended and rejected (see Klein, 1988, 2000). IOR appears when there is no endogenous motivation to move attention voluntarily to the cued object/location. Except on rare occasions IOR is observed only in exogenous cuing paradigms where the cues do not predict the location of an upcoming target. An exception is when allocating controlled attention to a location becomes advantageous (e.g.., when discrimination is difficult). In such a case, IOR can be overcome or at least substantially reduced by

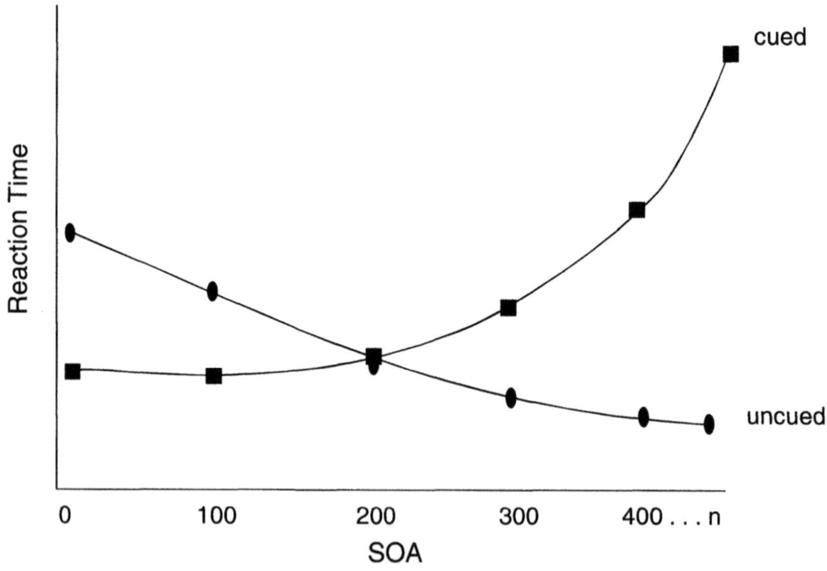

FIGURE 4.4. In a Posner (1980) cuing study with nonpredictive cues, the normal facilitation at the cued location changes to a cost as the onset time between cue and target increases. This cost (response time to cued vs. uncued locations) is known as inhibition of return, or IOR.

voluntarily keeping attention at the cued location (see Taylor & Klein, 1998). This finding does not take away from the reflexive nature of exogenous orienting. Effort can reduce reflexive actions; even a knee jerk induced by a physician's hammer can be reduced by cognitive effort.

Tipper, Driver, & Weaver (1991) and Tipper (1994) used exogenous cues in their studies and examined IOR at both the cued location where an object had been and the location to which the cued object moved. They found that IOR was present at both, and concluded that IOR is both space-based and object-based.

Frame Dragging

However, another way that these results could be obtained would be if attention were allocated within a spatial frame that had rotated around the origin of fixation in concordance with the movement of the boxes. If the left box were cued, that box would remain the left box within the rotated frame, but what would happen if the boxes moved in such a way that they broke the frame during movement? The boxes would still be objects in the Tipper sense, but their spatial relationship would be broken. Krista Schendel (2001) and I addressed this question by examining IOR in cases where the boxes moved as in Tipper et al.'s experiments compared to cases when the boxes moved away and then toward each other through a corner angle or in opposite directions (Figure 4.5). Notice that in each case the objects ended up at the same locations and only their path of motion changed. When the motion ceased, the target appeared either in the cued box or in the uncued box and subjects responded by a key press when they detected the target. Eye movements were monitored to ensure that subjects fixated centrally during the entire trial.

IOR was only observed when the boxes moved in a manner that was consistent with a frame rotation. It disappeared when the frame could not be used to drag the objects and thus their locations along with it. These findings are consistent with frame rotation, something like that shown in Figure 4.6. When the two objects in the field maintained their spatial unity, the spatial referents could be maintained. When they did not, the spatial referents were abolished, and the "object-based effects" disappeared (also see Christ, McCrae, & Abrams, 2002).

One could, of course, argue that common fate in the rotating condition grouped the objects together, and it was this grouping that maintained the IOR effects, not the frame of reference per se. In this way the two boxes became one object, so it could be argued that IOR in this sense was object-based. But this would miss the point. Grouping allowed the spatial referents within the display to survive rotation, but it was the spatial ref-

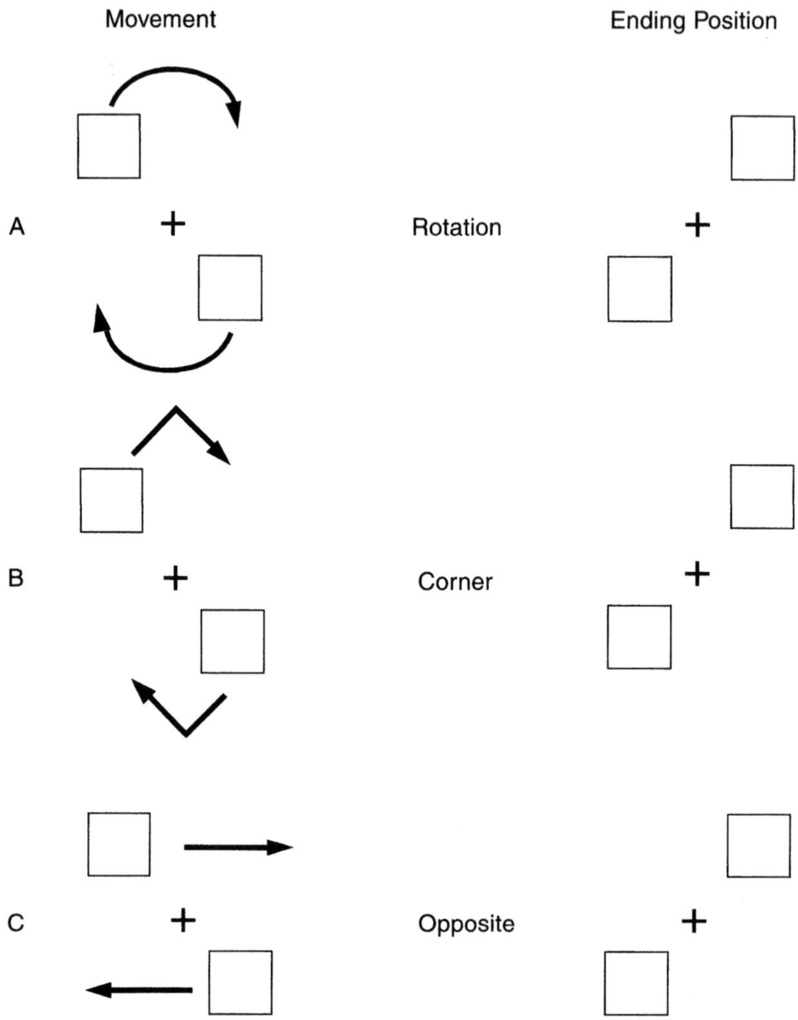

FIGURE 4.5. Manipulation of the path of motion in the moving boxes experiment represented in Figure 4.3. The boxes moved together through rotation (a), or moved in separate directions either by turning a 90° corner (b) or by passing each other vertically or horizontally in opposite directions (c). (From Schendel & Robertson, 2000.)

erents (left and right in the reference frame) that defined the position of the two boxes and accounted for the attentional effects. A cued left box remained the left box and continued to be inhibited at longer SOAs, and an uncued right box remained the right box and was not.

In fact, there was no evidence that grouping through common fate produced any inhibition of the uncued box at all, as would be predicted if

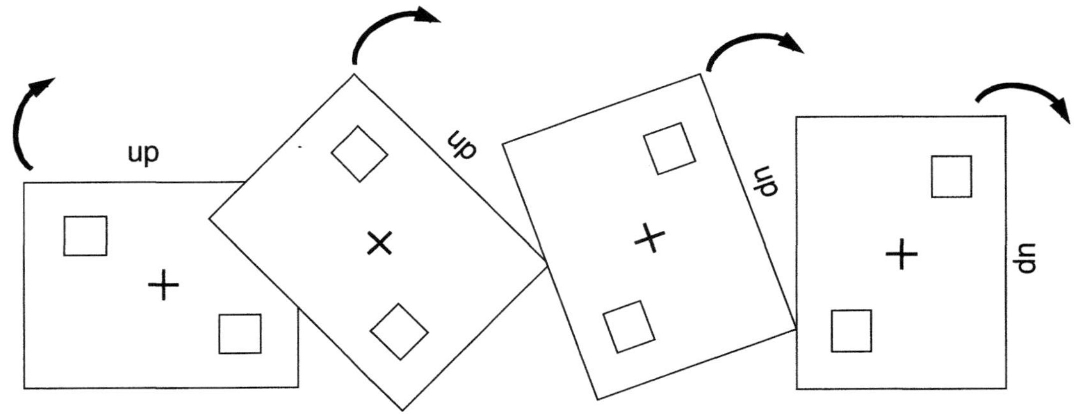

FIGURE 4.6. Example of how two boxes that have been defined as different objects in the literature maintain their relative spatial positions by a rotation of a spatial frame.

IOR were directed toward the entire object group. Reaction time to detect a target in the uncued box was not significantly different across the three conditions (Figure 4.7). These data suggest that it was the spatial referents of the group that determined IOR and not grouping through common fate that best accounted for the results. It is the frame that appears to rotate, dragging the boxes and their history (which one was cued) along with it. IOR appears to be space-based in an updated frame of reference.

This argument can be extended to include other cases of object-based IOR, such as that reported by Gibson and Egeth (1994). They cued a location within a brick-like stimulus and then rotated the brick in depth before presenting a target. When the target was in the same relative position on the brick, IOR was observed. It was also observed when it was in the same position in environmental coordinates. Again, without maintaining the spatial referents of the object, we would expect IOR within the brick to disappear but to be observed within the environment that remains stationary.

As discussed in chapter 3, endogenous or predictive cuing is also sensitive to rotating frames, but in these cases natural, baseline directional biases were used to study the influence of reference frames on spatial attention (Rhodes & Robertson, 2002; Robertson, 1996). Recall that endogenous cues do not produce IOR, so the effects are facilitory even at long SOAs. Although we did not examine the effects of endogenous cuing in the rotating boxes experiment, using such a procedure would not address the question of frame-dragging in endogenous cuing. It is doubt-

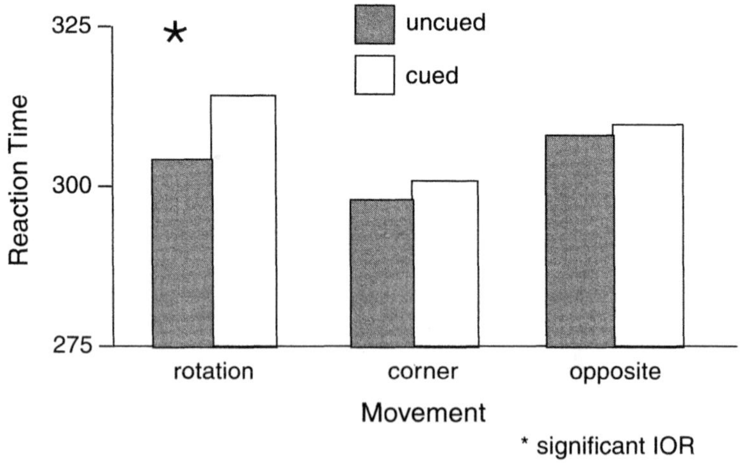

FIGURE 4.7. Mean reaction time to detect a target in the cued versus uncued box in the conditions represented in Figure 4.5. Only the rotation condition produced significant IOR (uncued faster than cued). (From Schendel & Robertson, 2000.)

ful that cues would disrupt facilitation in the different conditions of Figure 4.5, as the visual tracking literature has demonstrated that individual items that are endogenously cued can be attended even through much more complex movements than those used in our studies (Pylyshyn & Storm, 1988). This literature has shown that subjects can successfully track from three to seven targets that randomly move in a visually cluttered display and this tracking facilitates response.

Frame Disruptions by Visual Tracking

Visual tracking studies generally include many randomly placed items on a computer screen. A subsection of these items is cued by something like brightening for a brief period of time, and then all the items on the screen begin to move simultaneously but in different paths of motion. When the motion stops, subjects are asked to locate the items they were supposed to track. In visual tracking studies the referent locations between points is broken and thus the reference frame that could guide attention in space (and presumably increase the number of items tracked) is also broken or at the very least ambiguous. Although a single dot is unlikely to contain its own frame (and in this sense may in fact be a pure measure of what has been called object-based attention), a rectangle surrounding the items to be tracked would contain such a frame. Attention could track an object in a space that defines a particular level of structure and does not move itself.

Yantis (1992) reported an interesting variant of the visual tracking procedure in which the targets could form groups so that their spatial referents to each other could be maintained. When this occurred the number of items that could be tracked increased significantly. It would be interesting to know how many groups can be tracked at any given time, but in any case, this is an issue for endogenous cuing.

Because the visual tracking literature demonstrates that attention can track targets moving through random spatial paths when the subject is motivated to do so, both Logan and Rhodes and I used standing spatial biases to evaluate endogenous or controlled attentional effects in reference frames (Logan, 1996, Robertson, 1996; Rhodes & Robertson, 2002), as described in chapter 3. However, at the very least, the visual tracking literature suggests that endogenous cuing of a box could motivate subjects to track a particular item through its path. I am not suggesting that following an object with attention cannot be done, but the following is through some space. When attention has to keep track of the locations of more than seven items, it breaks down. It is not only the number of items but also the number of locations (more specifically, spatial paths) that may contribute to visual tracking limitations.

In sum, the findings using rotation and exogenous cuing to examine object- and location-based attention can be explained by spatial attention that is allocated within spatial reference frames. The data discussed in this section demonstrate that at least one measure that has been used to study attentional orienting (IOR) can be attributed to the spatial referents in these frames.

The rotation studies strongly suggested that IOR was maintained within a spatially updated frame. When an object location is defined, whether by a more global object (Gibson & Egeth, 1994) or by common motion (Schendel 2001), the spatial referents within the frame are maintained, and thus attention to locations within that space are maintained. When two boxes are grouped through common fate, the frame's origin can be centered on fixation and the items in the group can maintain their spatial position through frame rotation and visual tracking.

Object-Based IOR and Illusory Contours in Static Displays

"Frame dragging" could account for object-based IOR in rotating frames, but there are other reports of object-based IOR in static displays. In one, a set of "Pacmen" was arranged in such a way that a subset produced illusory contours forming a square shape (Figure 4.8). The question was

FIGURE 4.8. Example of "Pacman" figures placed such that illusory contours form two squares, one to the right and one to the left of center. (Adapted from Jordon & Tipper, 1998.)

whether cuing effects would be stronger when the Pacmen formed illusory contours that looked like square boxes than when they did not. The illusory contour shapes appeared either to the right and left or above and below fixation, as in the traditional Posner cuing paradigm, but with long SOAs between cue and target in order to produce IOR. Whether the Pacmen formed a square or no shape was varied (Jordan & Tipper, 1998).

On each trial, one of the locations where the squares could be centered was cued, and a target appeared at either the cued location or an uncued location with equal probability. IOR was present whether the Pacmen formed an illusory square or not, but there was significantly more IOR when they formed a square. Although it is possible that more inhibition accrued at the cued location when it was inhabited by something we call an object (i.e., a square shape as opposed to a location between randomly oriented Pacmen), it is also possible that the illusory contours defined a spatial location with more precision than the randomly placed Pacmen. Illusory contours form an area that is perceived as brighter than the background and that pull attention to these locations (Palmer, 1999). This would basically highlight the location of the illusory square as well as reduce spatial uncertainty.

In another study Jordan and Tipper (1999) also examined IOR using illusory contours, but in this case addressed the question of whether IOR would spread within an object. In this case objects were defined by two rectangles modeled after the stimuli used by Egly, Driver, and Rafal (1994). The two rectangles were arranged so that the cued location (one end of one of the rectangles) was equidistant from an uncued location within the cued object and an uncued location within the uncued object (left rectangles in Figure 4.9). Endogenous cuing generally produces strong object-based benefits in detection. Reaction times to detect targets at cued locations are faster than at uncued locations within the same object and at uncued locations in a different object. This is not the case for IOR. IOR was present at the cued location as expected in Jordan and Tipper's (1998) study, but it did spread differentially within and between objects. Nevertheless, in rectangles created from illusory contours (right figure in Figure 4.9), significant IOR was not present. In sum, the role of objects in producing IOR is somewhat equivocal as is the role illusory contours play.

Studies performed by Alexandra List (a.k.a. Alexandra Beuche) and myself (2000) have gone on to show that IOR is specific to the cued location in stimuli identical to those used by Egly, Driver, and Rafal (1994) (Robertson & Beuche, 2000). Not only was there no evidence that IOR spread within cued objects, but in fact the opposite occurred. A benefit was observed at the uncued location within the cued object relative to the uncued location between objects at the same time that IOR appears at the cued location as expected. This finding will require a bit of explaining, so I will begin with details of the experimental methods.

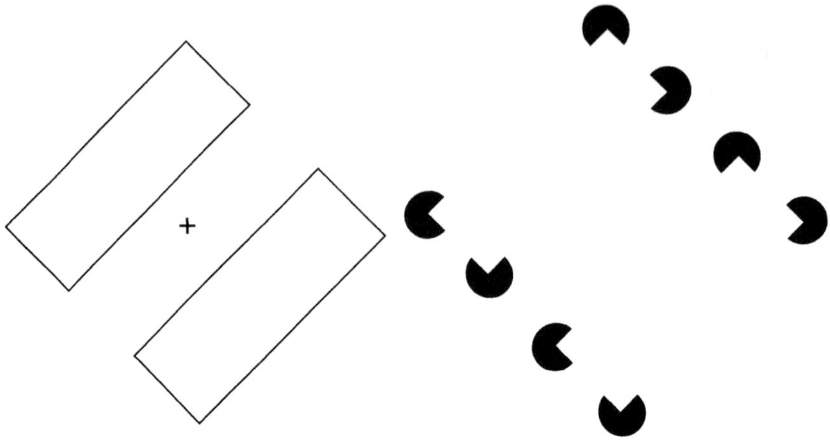

FIGURE 4.9. The two rectangles on the left are similar to the stimuli used in a cuing study to examine object-based attention by Egly, Driver, and Rafal, 1994, while the two rectangles on the right are defined by illusory contours and are similar to those use by Jordan and Tipper (1998).

We presented a pair of rectangles like those used by Egly, Driver, and Rafal (1994) on each trial (rectangles on the left of Figure 4.9 except vertically oriented). A cue appeared for 100 ms at the end of one of the rectangles on each trial, and a target was presented either 300 or 950 ms after cue onset (SOA). Recall that Egly, Driver, and Rafal used a predictive, endogenous cue and found a response benefit at the cued location at 300 SOA (which we first replicated), but in another experiment we used nonpredictive, exogenous cues, and we found IOR at both 300 and 950 SOA. But more importantly, there was no hint that reflexive orienting as marked by IOR was sensitive to objects. If anything, the cue *benefited* detection within the cued object compared to the uncued object. In other words, despite the unpredictive nature of the cue, which was clearly effective in producing IOR, the object-based effects (within- vs. between-object differences at uncued locations) was the same as that found by Egly, Driver, and Rafal and opposite that found by Jordan and Tipper (1999). Target detection at uncued locations within the cued object benefited response time relative to an equally distant target in the uncued object.

In a recent paper Leek, Reppa, and Tipper (2003) defined object-based IOR in a different way, namely as the difference in reaction time to detect a target when rectangles were in the stimulus compared to when they were not. The authors argued that the slower detection time they observed when "objects" were present supported object-based IOR. But this effect could also mean that detection time is simply slowed in the presence of contours. A second finding that was interpreted as support for object-based IOR was in fact more consistent with object-based facilitation as List and I (2000) reported. When targets were presented within the same part but at an uncued location, detection time was faster than when they were presented in a different part at equal eccentricities and at equal distances from the cued location.

Perhaps it is time to back up just a bit and go through the Egly, Driver, and Rafal (1994) method and their findings in more detail to understand what all this might mean, especially because their methods have been used so often to study object-based attention. As I just mentioned, they used predictive cues (endogenous cuing) and found a benefit at both the cued location and an uncued location within the cued object compared to the uncued object (Figure 4.10). On each trial a peripheral cue appeared for 100 ms (the graying of the outline of one of the ends of a rectangle, randomly determined). The cue informed the subject that a target would appear at the cued location 75% of the time. On the remaining trails, the target appeared equally often at the uncued location within the cued rectangle (within-object condition) and at the uncued location equidistant from the cue in the uncued rectangle (between-object condition). The target appeared 200 ms after cue offset, and participants were instructed to press a key when they detected it. A small number of catch trials were included in which no target appeared, and participants were instructed to withhold response on those trials. Catch trials were included to attenuate early responses, and they successfully did so.

The results of Egly, Driver, and Rafal's (1994) study showed that the effects of cuing were strongest at the cued location. Predictive cuing decreased the time it took to detect a target at that location as usual. More importantly, the object manipulation affected detection time. Despite the fact that the locations of uncued objects were the same distance from the cue in the display, subjects were faster to detect targets in the within-object condition than in the between-object condition (Figure 4.10). The magnitude of this effect was relatively small (13 ms), but it was very reliable and it has been replicated many times (see Egeth & Yantis, 1997).

The findings show that the configuration of the stimulus affects either the movement of endogenous attention from one location to another or the spread of attention over the visual field (i.e., a spatial gradient) in a way that is sensitive to object boundaries. In addition, this design elegantly overcame a major hurdle that was inherent in studies of object-based effects reported before it, namely that objects inhabit different locations. By examining performance at locations that were not the cued location but either in the same or a different object, this confound was eliminated.

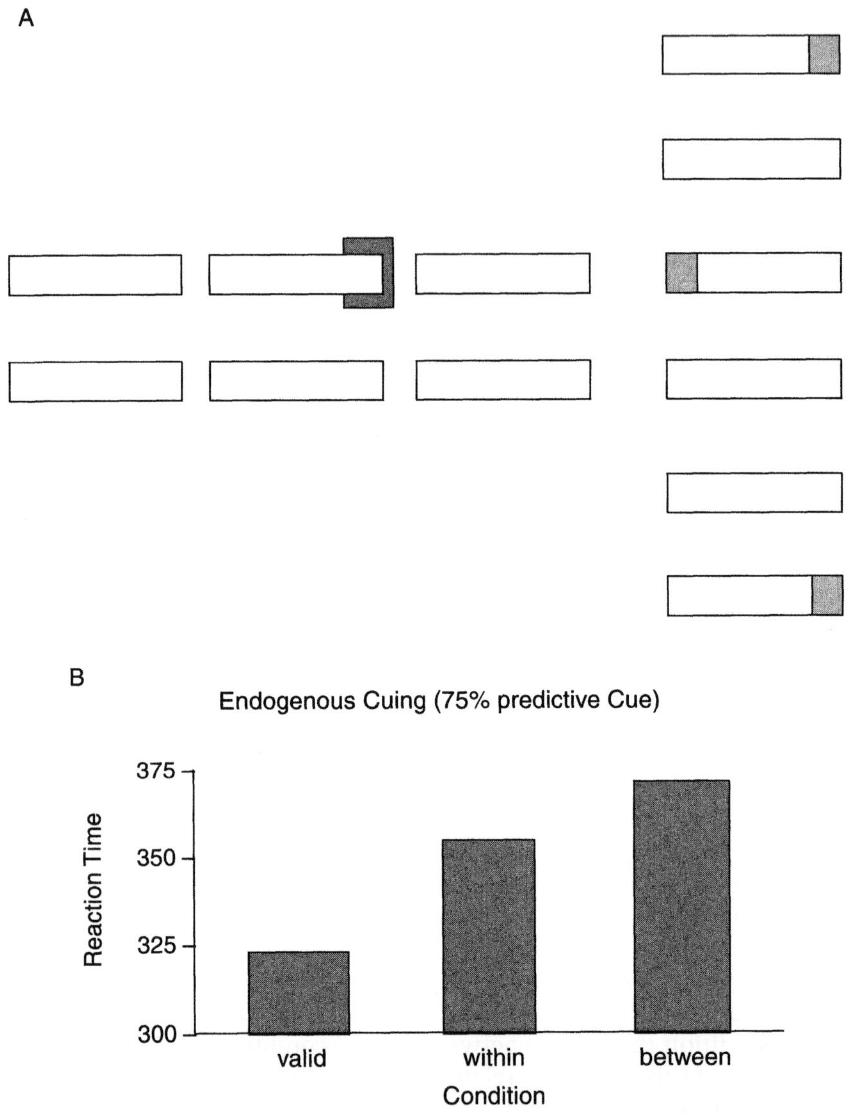

FIGURE 4.10. Example of a trial sequence in the study by Egly, Driver, and Rafal (1994), that examined object-based attention (a). Two rectangles (objects) appeared on the screen and 100 ms later one end of one of the objects was cued for 100 ms, informing the participant that a target was about to appear there, which it did on 75% of the trials. Two hundred ms later the target appeared at either the cued location (valid), an uncued location within the cued object (within), or an uncued location within the uncued object (between). Mean response time for validly cued locations was fastest, but within-object response times were faster than between-object response times (b). The difference between within and between conditions is the object-based effect. (Adapated from Egly, Driver, and Rafal, 1994).

Endogenous spatial attentional orienting and its resulting benefits on detection were sensitive to objects.

Space-Based Inhibition and Object- and Space-Based Facilitation

We are now in a position to return to the results obtained with exogenous cuing using the same stimulus displays as Egly, Driver, and Rafal (1994). Recall that we found IOR at the cued location as expected (response times in the valid condition were slow), but we also found that for uncued locations, within-object target detection was faster than between-object detection. Object-based IOR predicts the opposite effect. Slowing of response time (costs) should be strongest at the cued location, then at the uncued location within the cued object; it should be weakest in the uncued object. The response time pattern should have been the inverse of that found by Egly and others, who reported object-based benefits with endogenous cues. Instead, not only was there no object-based cost in a study where IOR was clearly present at the cued location, but there were actually object-based benefits.

How might this asymmetry between the effect of object boundaries on spatial costs and benefits be resolved? If benefits at cued locations are due to one spatial attentional mechanism, and costs to another, then their independent effects in the Egly task would not be particularly surprising. Our results suggest that benefits reflect sensitivity to the perceptual organization of the stimulus, but costs do not. Costs or IOR appear to be location-based and blind to objects, while facilitation is sensitive to object structure but also to space that defines that structure. Theoretically, IOR emerges later than facilitation, but our results suggest that facilitation can remain active at long SOAs at the cued location but that inhibition masks it in the response. Figure 4.11a shows a theoretical distribution of endogenous spatial attention over the Egly display shortly after a cue appears, while Figure 4.11b shows the location-specific inhibition that can occur early or late but is almost always present at long SOAs with unpredictive cues. Figure 4.11c shows what would happen if the two attentional effects were placed on top of each other. Location-based inhibition would produce IOR at the cued location while facilitation would produce a within-object over between-object advantage, and this is what we found (Figure 4.12). Inhibition does not follow or replace facilitation after a cue. Rather, both appear to operate in parallel to influence the overall pattern of results (see Klein, 2000).

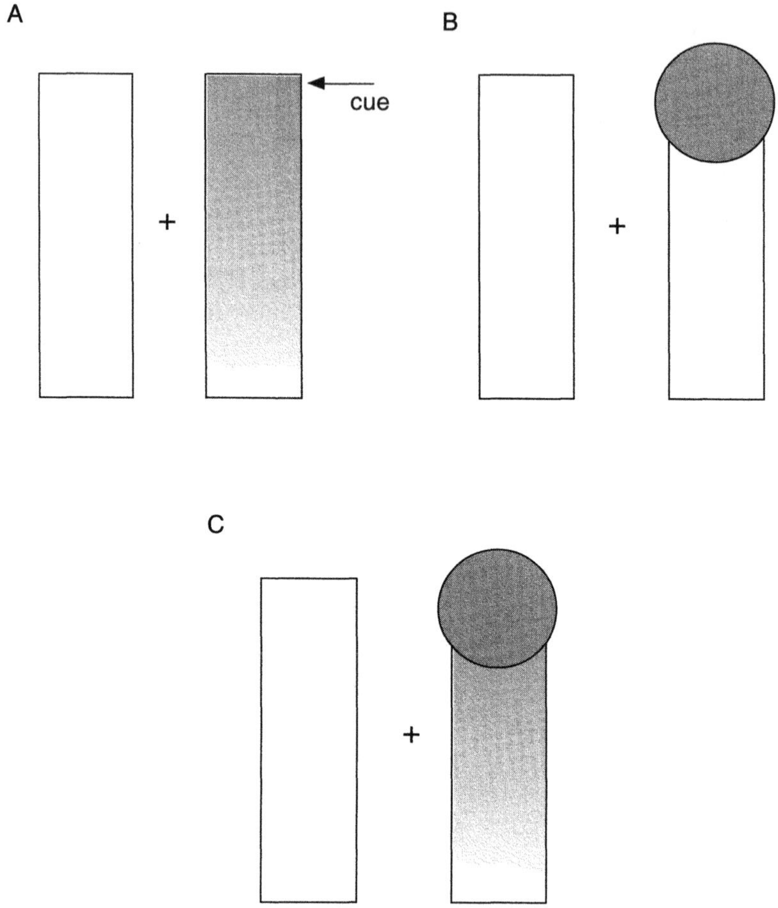

FIGURE 4.11. A theoretical distribution of two attentional systems acting together: one that produces benefits that are sensitive to perceptual organization of the display (a) and one that produces inhibition that is only sensitive to space (b). When the two are superimposed (c), there will be costs at the cued location when b is stronger than a, while uncued locations within the same object will continue to benefit relative to uncued locations in the uncued object.

Object-Based Effects and Perceptual Organization

The attention literature tends to discuss objects as if everyone knows what an object is, but it seems to be whatever the experimenters call an object in any given study. An object can be a rectangle, a flower, a column of letters, the head of a pin, a forest—anything that appears as one perceptual unit as opposed to another. The slippery nature of objects was driven home to me when Min-Shik Kim and I (Kim & Robertson, 2001) asked

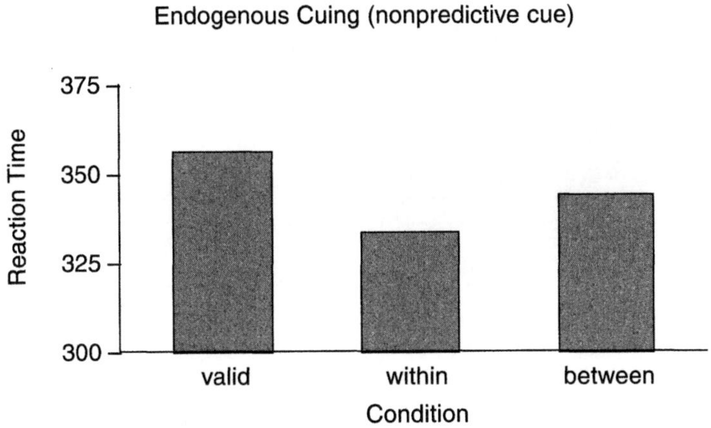

FIGURE 4.12. Mean response times were overall slower for valid conditions, consistent with location-based IOR, but were still faster in the within than between condition.

the question of how perceived space (as opposed to space measured by a ruler) would affect the object-based effects reported by others. In order to address this question we placed two black rectangles (a modification of the Egly, Driver, and Rafal, 1994, stimulus) in the context of a room that created a spatial illusion (Figure 4.13). Although the two dark lines look vastly different in length, they are in fact the same, and the distance between them is the same as this length. This illusion was first published by Rock in 1984 as a real-world example of the Muller-Lyer illusion, but we changed the parameters to accommodate the Egly stimuli. The question we asked was whether attention was allocated within space as it is projected to the visual system (e.g., retina and early visual cortex) or to space as it is perceived. The answer was space as it is perceived.

By using the "room illusion" and the same methods as those used by Egly, Driver, and Rafal (1994), we first demonstrated that responses to invalid target locations within the cued line were slower when the perceived distance between the cue and target was longer than when it was shorter. In other words, it was the perceived line length that determined the spread of spatial attention within the cued line, not the distance on the screen. We also replicated the object-based effects reported by Egly, Driver, and Rafal. Responses to invalid target locations within the cued line were faster than to invalid locations in the uncued line.

When we first presented these results, I suggested that the object-based Egly effect might actually be due to a perceived distance effect (i.e., the distance between the two lines appears larger due to depth cues in the room illusion). Several people took issue with this conclusion and pointed out that the lines in the stimulus could be considered objects. Although

FIGURE 4.13. Example of the "room illusion" stimuli. The two vertical dark rectangular lines are perceived as different sizes, with the one on the back corner perceived as longer than the one on the front. However, their heights are the same and the distance between them is the same as their heights when measured by a ruler. (Adapted from Rock, 1984.)

we can conceive of the two dark lines as objects if so inclined, it seemed rather arbitrary to call the corners of a room objects, especially a portion of a corner of a room (the dark line in the foreground). If these were objects, then what wasn't an object?

In order to understand the importance of these results, a more thorough discussion might be useful for those who remain unconvinced. We used the same timing procedures and cue predictability as Egly, Driver, and Rafal (1994). We changed the cue to a red bracket that marked one end of one of the dark lines for 100 ms. The target (a small white dot positioned just within the borders of the dark line) appeared 200 ms later, and subjects responded with a key press when they detected the target. Catch trials were included as well in which no target appeared and responses were to be withheld.

As noted above, the first question was whether the perceived line length would influence detection time, and it did. We also obtained a normal Posner cuing effect with targets at cued locations detected faster than at other locations. Importantly, reaction time did not differ for target detection when comparing only validly cued locations in the perceived longer and perceived shorter lines, demonstrating that local stimulus factors that differed around the ends of the two lines did not affect target detection. For instance, the perceived longer line ends at the ceiling with the lines designating the connecting walls close by. The equal RT in the cued conditions reduces concerns about any differential masking effects that could have accounted for the results.

In another set of studies (Barnes, Beuche, & Robertson, 1999) we examined the influence of the illusion on the pattern of inhibition or IOR. I have already discussed findings that suggest that IOR is space-based, but does it respond to perceived space? By using the room illusion stimuli once again, we could determine whether the results supporting location specificity of IOR would also be present in a more complex scene. The SOAs we used were 300 and 950 ms, but the cue was not predictive. Indeed, IOR was present but it was not influenced by the perceived space within the illusion. We found no evidence in three separate studies that IOR was affected by perceived space, at least not the perceived space of the illusion shown in Figure 4.13.

Again, the Egly object-based effect was present, but as we previously found with simple rectangles (Figure 4.12), there was no evidence that IOR was sensitive to objects. Instead, responses to detect targets in the within-line conditions were faster than in the between-line conditions. In other words, a benefit for the uncued locations within the cued object was still present. Again, these data support a combined space-specific inhibitory effect that is present in parallel with an object-sensitive facilitory effect. These combined influences on spatial cuing were also found in static displays in a different study in which the target either appeared in a new location or changed to a new object (in the same location). Early benefits were sensitive to objects, while later IOR was not (Schendel, Robertson, & Treisman, 2001).

Basis of Exogenous Spatial Orienting and Inhibition of Return

There is a great deal of evidence that spatial orienting to an abrupt onset is reflexive and engages neural systems involved in saccadic eye movement programming (see Klein, 2000; Rafal & Robertson, 1995). In fact, there has been a long history of relating covert spatial attention effects to motor responses or preparation for action (e.g., Rizzolatti, Riggio, & Sheliga, 1994); fMRI studies have shown a remarkable correlation between cortical areas involved in eye movements and covert attention (Corbetta et al., 1998).

In a Posner (1980) cuing study, subjects are instructed to fixate one point (typically a central fixation pattern) and attend to another (generally to a cued location), creating a case where eye movements that might be reflexively generated under naturally occurring conditions would need to be inhibited. An eye movement itself is clearly not necessary to attend to locations in the periphery, and several studies have monitored eye movements to verify that such movements cannot account for spatial cuing effects. However, this does not mean that the computations that normally occur in oculomotor planning have not been performed. The plan could be present without implementation of the plan, and no amount of eye movement monitoring can help in determining when the plan is initiated and when it is not. This in essence is the premotor theory of attention (Rizzolatti et al., 1994).

Spatial attention and eye movement responses are often tightly coupled, and it would seem beneficial to have evolved a system that automatically orients to abrupt onsets, since these often signal potentially threatening information. On the other hand, not every stimulus is a stimulus that would benefit from automatic orienting. Certainly this is the case when considering manual responses. Automatically reaching for an object that suddenly appears in the periphery (e.g., a lit match or a snake) could have bad consequences, and it is clear that mechanisms of orienting have not evolved to make an automatic manual response to every stimulus that appears. This may seem like a trivial statement until one realizes that we can make the same arguments for oculomotor responses. Although we do not get burned or bitten by moving our eyes to a peripheral stimulus, there are conditions where eye movements toward a stimulus are not innocuous. For instance, orienting to an extremely bright light can cause eye damage, and orienting to a projectile on course with the eye would be very counterproductive. For some animals, eye contact is a sign of aggression, and diverting the eyes in another direction is a sign of submission. It therefore seems that an attentive preview of the stimulus would be prudent before an eye movement is planned and made. A space-mediated system that previews objects in particular locations would seem extremely beneficial.

In addition, after a saccadic eye movement has been made, the location is tagged in a way that inhibits the return of fixation to the tagged location after another eye movement. It has been suggested that this mechanism motivates attentional exploration by biasing eye movements to locations that have not already been sampled (Posner, Rafal, Choate, &

Vaughn, 1985). Theoretically, this function is thought to be the basis for IOR (Klein, 1988).

Although the cuing paradigm with its elegant simplicity has had a dramatic effect on attentional theories, it is clear that not all visual cues are made alike. Even the simple presentation of a peripheral cue can produce very different effects depending on the task, stimulus parameters and whether particular brain structures are intact or not. Over 20 years of research using Posner cuing paradigms has provided very good evidence that peripheral cues automatically activate midbrain structures that govern saccadic eye movements (the superior colliculus, or SC). However, a peripheral cue alone is clearly not sufficient to induce a saccade. We can choose not to move our eyes to bright flashes of light in the periphery, but whether we do or not, the same cells in the SC will fire as if we had prepared to make an eye movement to the cued location (Dorris, Taylor, Klein, & Munoz, 1999). This correspondence suggests that some type of inhibitory signal that cancels saccades are sent to the cells in question. This inhibitory signal appears to come from the frontal eye fields (FEF), which are strongly connected to SC. Henik, Rafal, and Rhodes (1994) demonstrated in neurological patients that a unilateral lesion in the FEF disinhibited saccades into the contralesional visual field. These patients were actually faster to make reflexive saccades to peripheral cues that were presented in the field that projects to their damaged hemisphere (i.e., the half of the field that should therefore be most affected).

As mentioned before, it has long been known that an eye movement to a location and then to another location will inhibit the latency of moving the eyes back to where they have just been (inhibition of return in saccades). At first glance, attention appears to be subject to the same rules. However, IOR is not present when central cues are used to direct attention *unless* saccade preparation is part of the experimental task (Rafal, Calabresi, Brennan, & Sciolto, 1989), and IOR is not necessarily present with peripheral cues (e.g., when the cues have predictive value). In this case peripheral cues act like central cues. They produce facilitation across short and long SOAs, suggesting that controlled, voluntary attention can basically cancel or ignore plans for reflexive orienting. To the extent that IOR is a signature of SC involvement in saccadic preparation, this effect would suggest that a separate mechanism is involved in voluntarily or endogenously allocating spatial attention.

There are several other converging bits of evidence that the SC is a critical structure in producing IOR. Some years ago Rafal, Posner, Freidman, Inhoff, and Bernstein (1988) demonstrated that IOR was reduced or eliminated along the same axes that eye movements were affected in a degenerative disease known as progressive supranuclear palsy (PSP). This neurological disease affects midbrain and frontal areas, and in the early stages, eye movements are impaired mainly along the vertical axis which later spreads to horizontal. IOR in this population was shown to be affected along the same axis as eye movements. A more recent report by Saper, Soroker, Berger, and Henik (1999) in a rare single case study of a patient with unilateral SC damage confirmed that SC is a critical structure in disrupting IOR.

An additional piece of evidence was reported in normal subjects by exploiting the differential representation in the SC for temporal and nasal sides of each eye (Rafal et al., 1989). Right and left visual fields are represented separately in the visual cortex both under monocular and binocular conditions. Information shown to the right side of each eye projects directly to the left visual cortex, and information shown to the left side of each eye projects directly to the right visual cortex. However, the relationship between right and left visual fields and SC afferents is quite different. The temporal (outer) sides of each eye are more strongly represented in the SC than the nasal sides (inner). With a design that examined IOR with temporal versus nasal cuing (monocularly), Rafal et al. (1989) demonstrated that IOR was larger in temporal than inn nasal spatial locations. IOR was larger in areas that projected more strongly to the SC. More recently, Berger and Henik (2000) have shown that IOR reduction by endogenous or voluntary attentional allocation is limited to nasal hemifields where IOR is not as strong to begin with.

Finally, Danziger, Fendrich, and Rafal (1997) showed that IOR was present in a neurological patient with primary visual cortex infarction, producing a homonymous hemianopia (blindness in the contralesional field). In other words, even when no visual information could be registered through primary visual cortex (V1), IOR was still present in both visual fields, presumably because the SC was intact.

In sum, the behavioral and neurobiological evidence together suggest that IOR is a marker for exogenous attentional orienting which is likely linked to oculomotor programming to different spatial locations. However, the space for this programming can occur in selected spatial frames and need not be limited to retinal spatial coordinates.

Object-based Effects and Perceptual Organization Revisited

The evidence that the SC/FEF oculomotor programming functions are involved in exogenous or reflexive orienting is quite convincing. But my discussion has been something of a diversion in order to come back to the question of how best to interpret evidence for object-based effects in IOR. Evolutionarily speaking, the SC is a very old part of the brain that is integrally involved in the generation of saccadic eye movements (and thus reflexive spatial orienting). Together the evidence for attention's link to saccadic inhibition of return and the evidence that it occurs in something other than retinal coordinates (Tipper et al., 1991) needs explanation. The SC is strongly connected to regions within both the parietal and frontal lobes that have been implicated in spatial orienting. The findings that IOR moves with a cued box during common motion shows that the SC is capable of either updating the spatial information in scene-based coordinates (Schendel, 2001; Rhodes & Robertson, 2002) or attending to objects (Tipper et al., 1994). The evidence against object-based IOR that List and I reported showed that IOR was specific to the cued location even in static displays, while facilitation continued to be influenced by the perceptual organization of the stimulus (see also Schendel et al., 2001). These results together suggest that spatial updating of a reference frame is a more likely scenario to account for IOR in moving displays. The cued object in the List and Robertson study showed no evidence of spreading inhibition within an object, and in fact demonstrated the reverse both with the original Egly-type stimuli and in the context of a room. In contrast to facilitation, IOR was not sensitive to the object or the perceptual organization of the scene. It was sensitive only to the cued location within the scene.

Object- and Space-Based Orienting and Exogenous Attention

What are we left with in terms of automatic spatial orienting and object-based attention? There is good evidence that exogenous spatial orienting is linked to a system involved in oculomotor planning. When IOR is evident, it signals that this system most likely has contributed to performance. It is clear that IOR is not limited to retinotopic locations (Posner & Cohen, 1984) and can move with the display as long as the display remains spatially coherent (Abrams 2002; Schendel, 2001). When elements (individual objects) move in such a way that the scene-based frame collapses (see Figure 4.5), IOR disappears. The evidence to support object-based IOR disappears as well. Instead the data as a whole become more parsimoniously interpreted as spatial inhibition within a selected spatial reference frame.

☐ Controlled Spatial Attention and Object-Based Effects

In chapter 3 I discussed at length the evidence that spatial frames can guide attention and produce facilitation at the cued location. There are also several studies that demonstrate very convincingly that attending to objects and/or their features can affect performance. For instance, negative priming effects, in which one shape is inhibited and another facilitated, show that objects and the attentional operations that were associated with them at the time of selection are represented over time (Allport et al., 1985), sometimes even for days (DeSchepper & Treisman, 1996). Conjoining features such as shape, color, texture, and size, seem to require attention (Treisman & Gelade, 1980; Treisman & Schmidt, 1982).

Representations of objects (whether or not we have good definitions of what they are or how they are represented) are clearly fundamental in everyday life. Approaching a tiger and approaching your spouse do not have the same consequences (at least under normal conditions), so knowing what an object is before acting would seem wise. Objects are of central importance, but objects do have a spatial structure. I have just argued that a spatial orienting system tied to oculomotor programming (that can be marked by the presence of IOR) responds to space within a selected frame. This frame may or may not be confined to what we call a single object, depending on which frame is selected. The mechanism underlying IOR seems to be a clear example of a space-based system, but the accumulation of evidence suggests that it is separate from another attentional system that is used for attentional control. To what extent are controlled attentional mechanisms object-based?

As I have argued throughout this book, it is likely that they are not strictly object-based but respond to space-based frames of reference that organize "objects" into hierarchical structures. Objects and space together define objects/spaces in which spatial attention can be allocated. The continual interaction between what and where systems produces a structured visual world, which is neither just objects nor just space. In such a world, we cannot select objects without accessing some type of spatial structure and we cannot attend to space within an object without spatial information. However, just as an exogenous spatial system that may be associated with midbrain structures can automatically represent a location for action (in this case for saccadic eye movements), so too could a system guided by principles of perceptual organization (e.g., grouping, closure, figure/ground, common fate, etc.) or familiarity (e.g., your name) automatically bring an object into awareness. Some attributes signal the presence of a new object for attention, one that replaces the old object of attention (Hillstrom & Yantis, 1994; Yantis & Hillstrom, 1994).

The neuropsychological literature also supports the automatic capture of attention by objects. For instance, both perceptual organization and unique features such as color affect what will be seen by patients with Balint's syndrome at any given moment (Humphreys, Cinel, Wolfe, Olson, & Klempen, 2000; Rafal, 1996; Robertson et al., 1997). Single objects seem to grab attention but then disappear as abruptly as they appeared.

Conversely, volitionally selecting an object for these patients is nearly impossible. There is no executive control over what object will be seen next. The stimulus flux in the visual world seems to automatically determine what will be seen and when.

Although Balint's syndrome is often heralded as a pure example of object-based *attention*, it is not an example of object-based *selection*. Recent evidence collected by Anne Treisman and myself show that once selection is required, whether of an object or of a spatial location either within or between objects, performance breaks down in these patients (Robertson & Treisman, in preparation). Given that temporal lobes remain intact (the ventral "what" processing stream), this syndrome seems to indicate that the temporal lobes themselves are not sufficient for selecting objects through attention, although they are sufficient for perceptual organization to occur and for single familiar objects to be formed (Humphreys et al., 2000). More will be said about Balint's syndrome and its implications for object and space perception in a later chapter, but the point here is that when considering attention as a controlled selection mechanism, damage to both parietal lobes appears to affect selective attention of both space and objects. Parietal deficits in attentional selection are not limited to the spatial realm.

Object-Based Effects and Endogenous Attention

The foregoing discussion has left out the question of how to understand the facilitory component of spatial orienting to objects and space. After all, the major studies (Duncan, 1984; Egly, Driver, & Rafal, 1994) focused on object-based benefits, not costs. Do within-object advantages, such as those observed in the Egly paradigm, represent a pure example of an object-based attentional system? The answer appears to be no, because if objects were selected without selecting their space as well, facilitation would be equal for all locations within the cued object, a point made most clearly by Vecara and Farah (1994). The example of object-based facilitation reported by Egly, Driver, and Rafal (1994), and replicated by several others including us, is clearly consistent with this point. Invariably, response time to detect a target in an uncued location within a cued object is slower than to detect a target at the cued location. Locations that define the object are not equally facilitated across the object as a whole. Objects are not selected without their space.

Nevertheless, there is a great deal of evidence pointing to two attentional mechanisms operating in parallel that produce facilitation, one space-based and the other object-based. The most common view of how space-based mechanisms operate is that they bias the movement of an "attentional spotlight" or the allocation of processing resources producing a spatial gradient. In either case, the Egly, Driver, and Rafal's results demonstrate that a space-based mechanism is needed even within an object. A more recent view of object-based effects is that locations within objects are given attentional priority for a serial scanning mechanism (Shomstein & Yantis, in press).

There is another possibility as well, one that suggests that spatial attention is biased within a spatial frame centered on a cued object. Faster responses to within-object over between-object locations (that are equidistant from a cue and fixation as measured by a ruler or by visual angle) are due to attention moving within a selected reference frame. Note that the Egly, Driver, and Rafal conclusions rely on the assumption that attention is directed in a single unitary space. But when one considers an object/space hierarchy, object-based and space-based effects are the same. For instance, in the Egly, Driver, and Rafal stimuli, each rectangle defines a local spatial frame (each origin centered at the center of the rectangle) and a more global spatial frame (the pair of rectangles with the origin centered at fixation). The cued "object" may cue selection of one of the local frames, and when the target does not appear within this frame, a new frame must be selected (with a more global frame centered on the pair of rectangles). Apparently, the selection of the new frame with respect to the old can influence how rapidly attention can be shifted (Vecara & Farah, 1994), further suggesting the relevance of both the local and global reference frames in attentional selection.

Object- or Frame-Based Selection?

One might argue that frame- and object-based selection are simply different words for the same thing, that there is no issue except semantics. But if this is the case, then the tie to neurobiology is not nearly as straightforward as it at first appears. However, there are ways to distinguish selecting on the bases of objects versus on the basis of space, and we have some preliminary but suggestive evidence that supports stronger frame-based than object-based models of selection even when endogenous attention is required. For reasons that I won't belabor here, we asked what would happen to the object-based effects in the Egly design when the two objects were more like thick black lines. The lines we used were the same ones that were the corners of the room in the room illusion that we employed in our previous studies (Figure 4.13), but with the context of the room omitted. The striped background left only the two thick black lines in the stimulus (Figure 4.14). We then performed the regular Egly cuing experiment. Peripheral cues were predictive and occurred at the end of

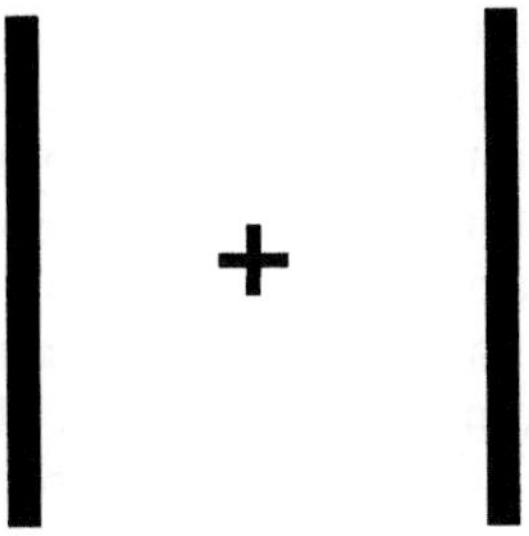

FIGURE 4.14. The two black vertical lines that were embedded in the room in Figure 4.13.

one of the lines, followed 200 ms later by a target at the cued location or at equidistant locations either within the cued line or the uncued line. As usually found with predictive cues, target detection at the cued location was faster than at uncued locations, but much to our surprise—and to that of many of our colleagues—there was no difference between within-line and between-line conditions for uncued target locations. When the "objects" were not defined by outlined rectangles or the context of room walls, the within/between-object differences disappeared. One of the most replicable findings in the object-based attention literature was not present.

Because of the incredulity (and a few bets) of my colleagues, we repeated this experiment 3 times, and in no case was there ever a within/between-object difference or even a trend in the expected direction (all $Fs < 1$). Nevertheless, when open rectangles were used, the object-based effects again appeared. It was not something strange about the procedure, equipment, or participants that eliminated the object-based effects with lines. It was the way in which the lines themselves were represented.

Is it possible that the effect disappeared with lines because the selected frame was now centered on the pair of lines (the origin of the frame was at fixation). A simple line may not engender its own object-based frame of reference. Upon presenting these findings to our colleagues, there were many responses that essentially boiled down to a line not being an object. It does not have closure, and an attentional mechanism that selects on the basis of objects would not select a line as an object. This is a legitimate counterargument to the frame hypothesis, but recall that the same lines did produce within/between line differences in the expected direction when placed in a context where they became the corners of a room (Robertson & Kim, 1998). It is difficult to explain our effects on the basis of object-based selection that selects the line as a different object in the case of the room illusion but not when the two lines are presented alone on the screen. The perceptual organization of the stimulus as a whole is obviously relevant for endogenous attention. Perceptual organization is in part defined by spatial reference frames (see Palmer, 1999), and these frames are used to guide attention. In the room illusion, we perceive the corners of the room as rather far apart (much further than a ruler would accurately measure). In fact, within the room as perceived, the space between the lines is larger than the space within them. Attention moves in spaces that are created from the lines, angles, planes, and other elements of the array within the selected frame of reference (in this case a 3-D one centered on the display, as opposed to the 2-D one that defines the distance between the lines).

So where does this leave us in terms of object-based attention? The idea that attention can facilitate object-based tasks as well as location-based tasks is not controversial, but the question is whether there are different mechanisms that select objects versus those that select space, and some fairly recent imaging data suggests that there are not. Yantis and his colleagues (Serences, Schwarebach, Courtney, Golay, & Yantis, 2001; Yantis et al., 2002; Yantis & Serences, 2002) have shown that the same frontal areas adjacent to the frontal eye fields and parietal areas in the superior parietal lobe (SPL) of humans is transiently active both when switching between stimulus streams in two different locations and when switching between streams of superimposed objects in the same location in the display. When attention is sustained, these areas do not sustain activity, while other "specialized" areas do (posterior inferior parietal lobes for location, medial fusiform gyrus for houses, and lateral fusiform gyrus for faces).

In the object-based study (Serences et al., 2001), faces and houses were superimposed on one another, with houses morphing into other houses at the same time that faces morphed into other faces. Subjects were given a cue to either switch attention between faces and houses (shift cue) or maintain attention on the stream they were already monitoring (hold cue). Their task was to detect a specific face or house within the attended (cued) stream. fMRI activity in different regions was then evaluated over epochs of time to determine what areas sustained activity after a hold cue and decreased activity after a shift cue versus those that showed transient activity after a shift cue but little activity after a hold cue. The former would indicate areas that continuously monitored a selected object category, while the latter would indicate attentional selection itself.

The data are consistent with a dorsal selector that interacts with ventral areas that represent object categories. The SPL was transiently active whenever shifts between object categories were made, whether from faces to houses or vice versa. This is the same area that shows transient activity when attention is shifted between right and left stimulus streams (now composed of multiple letters). The signal to switch attention, whether between objects or locations, comes from the same source within the human cortex.

For the hold cue, the profile was different. First, areas that are known to increase activity to place stimuli (Epstein & Kanwisher, 1998) showed sustained activity while maintaining attention on the place stream, and areas that increase activity to face stimuli (Kanwisher, McDermott, & Chun, 1997) showed sustained activity when maintaining attention on the face stream. However, frontal and parietal areas that were activated by a shift cue showed little to no increased activation after hold cues.

It is not surprising that paying attention to faces activates face areas, and paying attention to houses activates place areas. This has been shown before (O'Craven, Downing, & Kanwisher, 1999). What the experiments from Yantis's lab demonstrate is that switching between two streams in the same retinal location versus streams in different retinal locations activates the same dorsal cortical area. Equally important is that they do so transiently, as would be expected if these areas were the source of a signal that switched attention between specialized processing areas. It appears that the switch signal is the same for locations and objects. The Yantis studies elegantly show that this signal is generated by dorsal processing and received by specialized areas within the ventral pathway.

But does a task that cues switching attention between houses and faces in the same location take space out of the equation? Actually, it does so only in one reference frame, and that is the frame tied to the retina. People perceive faces and places that are merged with each other on a computer screen as overlaid or superimposed stimulus categories. In fact, this is how we describe them, and it is consistent with the bias of the visual system to impose a 3-D spatial structure on such stimuli. Houses and faces are not in the same location in perception. Rather, they appear in different frames, one behind the other.

I already discussed data demonstrating that exogenous attention is influenced by the 3-D percept generated by a 2-D pattern when I described the room illusion experiments (Fig. 4.13). The distance between locations in retinal space did not account for the data as well as the distance between locations in perceptual space did (Robertson & Kim, 1998). I would assume that the same principals hold for superimposed streams of morphed faces and houses. One stream is seen as behind or in front of the other, and switching between these streams requires a switch in the spatial frame that is selected for attentional monitoring. The contents of this frame activate different cortical areas, while the selection process itself, whether between objects or locations, appears to be a function of the same mechanism, one that seems to determine which frame of reference is important for the task at hand.

Object-Based Neglect

In chapter 3, I discussed several results generated from patients with unilateral visual neglect which demonstrated that neglect can occur within a spatial frame of reference. The findings are often referenced as cases of object-based neglect. The literature in this area typically assumes (sometimes implicitly) that every result that cannot be tied to the retina reflects an object-based effect of some sort (although see Mesulam, 1999, for an exception that includes multiple spatial frames).

A

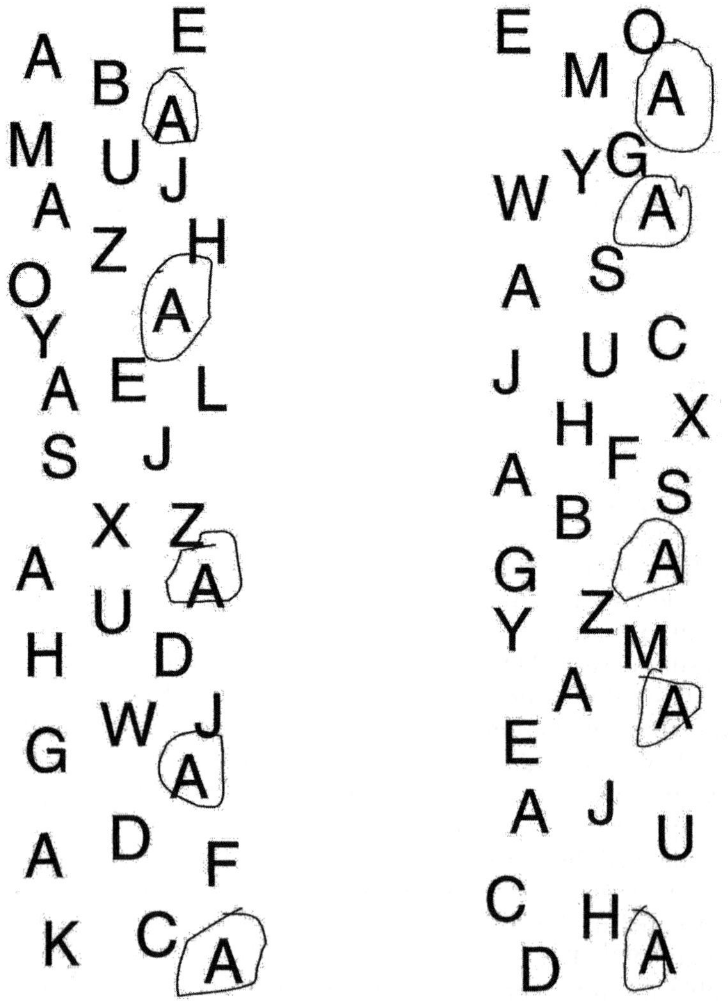

FIGURE 4.15. Examples of object-based neglect on a standard clinical test for neglect (see text). The patient was asked to circle all the As that were on the page (a) and to draw the upper figure in (b). The patient's drawing is shown below.

B

FIGURE 4.15b.

A simple example from a neuropsychological test known as the Standard Comprehensive Assessment of Neglect (SCAN) makes this point well. This test is an improvement over previous tests such that it includes an evaluation of both space- and object-based neglect. One example of object-based neglect from one of our patients is shown in Figure 4.15a. Note that the patient (with right hemisphere damage and left neglect) circled all the As on the right side of each column but missed As on the left side of each column. On another subtest the patient drew the right side of objects in a room even when they were on the left side of the page, but missed parts on the left side of individual objects (Figure 4.15b). This example is easily labeled as object-based, but what should be made of the example from the columns? The parameter that makes each column an object is grouping by proximity. That is, the letters in each column are clustered together within the space of the whole display. Neglect for the left sides of the columns and the left sides of objects when copying a drawing of a room seems more parsimoniously related to the spatial referents that are necessary to perceive where the parts fit within the whole.

One could argue that this distinction is just semantics, but if so, where does that leave space-based neglect? Does the concept disappear, and if it does, then why not call object-based neglect simply neglect? This would be to miss the point. Neglect (whether labeled object or not) occurs within frames of reference and appears to be a spatial problem within the frame of reference currently attended. The left side of the spatial map that defines the spatial referents for that frame has been affected, whether based within retinal, viewer, scene, or object-based coordinates. This seems like an explanation that can account for a wide range of findings in the literature. Reaction time measures reveal neglect in both display-centered, cluster-centered, and item-centered coordinates as well as viewer-centered ones.

In a recent study we tested 6 patients with right hemisphere damage and left neglect who demonstrated both space-based and object-based neglect on the SCAN in a study designed to examine the rotation of global and local frames of reference on the magnitude of neglect (Schendel, Fahy, & Robertson, 2002). Local frames at the center of the display did not reveal neglect, while global frames did. But what was most interesting for the concern at hand was the combination of viewer-based and global-item-based frames of reference in the pattern of performance.

Figure 4.16a shows examples of some of the stimuli we used. Patterns with a local and global item were shown on a computer screen, and on all trials either the global or local object rotated 90° out of alignment from its starting upright orientation. Figure 4.16a includes two global rotation conditions. A small green dot then appeared at one of four locations that were in one of the four quadrants of the visual field (Figure 4.16b), and the patients simply pressed the mouse key whenever the green dot appeared. Reaction time patterns supported the joint influence of viewer-centered and global-centered frames of reference. The results are shown in Figure 4.16c, where the additive effects of the two frames are evident. The worst performance was when the target appeared in a quadrant that was both left within the object and left of the viewer, while the best performance was when the target appeared in a quadrant that was both right within the object and right of the viewer. The other two cells were in between. Neglect was affected by two frames of reference, one defined by the viewer and another defined by the orientation of the global form (i.e., the global object).

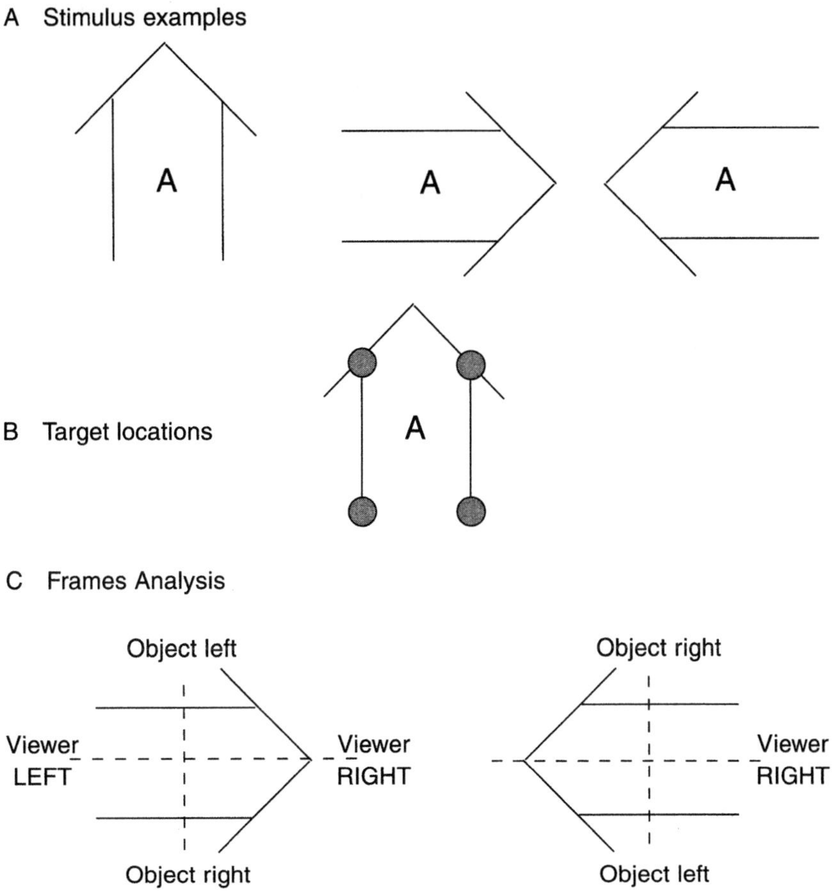

Mean detection time

		Viewer-based side: left	Viewer-based side: right
Object-based side	left	617	537
	right	592	529

FIGURE 4.16. Stimulus examples used to test a group of patients with left neglect (a). The global form rotated either 90° clockwise or 90°counterclockwise followed by a bright green target that appeared in one of four locations (b). Response times to detect the target were analyzed within object- and viewer-based left/right sides (c). Mean response times are presented at the bottom.

One of the most convincing bits of evidence that attention is allocated in multiple frames of reference comes from a study reported by Behrmann and Tipper (1999) with patients with left neglect. In a very clever design, a barbell was presented diagonally with one end of the barbell on the right and one on the left side of the display (Figure 4.17a). The barbell was situated between two unconnected boxes that were the same eccentricity from center but were unconnected. On each trial a target (a bright dot) appeared in one of the circles of the barbell or in one of the unconnected squares with equal probability. As one would expect, detection of a target in the left circle or square was significantly slower than in the right circle or square. This difference served as a baseline measure of neglect for the manipulation of interest, which was the rotation of the barbell so that the circle that was originally on the right moved to the left, and the one originally on the left moved to the right (Figure 4.17b). When the target appeared in the circle that was originally on the right (but now on the left, represented by the open circle in Figure 4.17b) it continued to be detected faster than when the target appeared in the circle that was originally on the left but now on the right (represented by the gray circle in Figure 4.17b). In other words, neglect was object-centered in the language of the field. The circle that was encoded as the right side remained coded as the right side even after rotation and benefited target detection despite the fact that it was now on the left side with respect to the viewer.

However, the most interesting finding was that the rotation of the barbell had no effect over baseline when the target appeared in the squares. Even when the right side of the barbell had rotated to the neglected side of the display, and was therefore closer to the square on the left, detecting

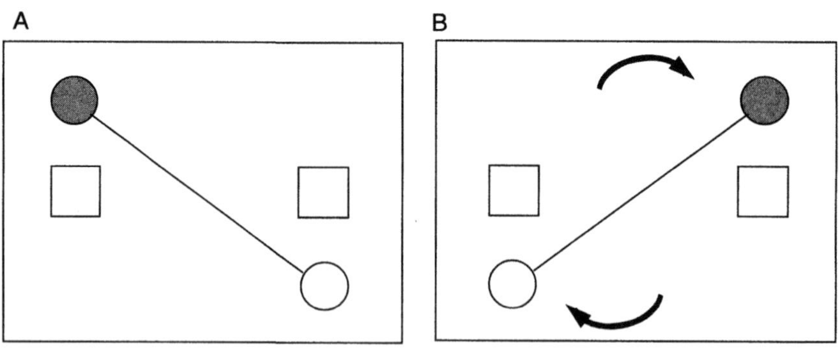

FIGURE 4.17. Stimulus examples used to test a group of patients with left neglect. A barbell and two squares appeared on the screen (a), and then the barbell rotated 90° clockwise or counterclockwise (b). After rotation a target appeared in one of the four shapes (either the circles on the barbell or the squares). (Adapted from Behrmann, & Tipper, 1999.)

targets presented in that square remained poor. Neglect of the left side in the frame of squares was dissociated from neglect of the left side within an object-based frame of reference.

It would be reasonable to conclude that attention tracked the right side of the rotating object (the barbell) during rotation, producing an object-based effect. However, this could not explain why the response pattern for targets in the stationary squares was the same for trials in which the barbell rotated and those in which it did not.

It appears that the barbell was represented in one spatial frame and the squares in another. The squares never moved and remained anchored to a frame that originally coded left as left and right as right in the display. Likewise, although the barbell rotated, its rotation was anchored to a frame that moved and maintained the spatial code of left as left and right as right within the barbell's intrinsic coordinate system. The internal representation of multiple spatial reference frames can account for what appears at first to be a paradoxical result. Attentional allocation in neglect was guided by the same principals in both frames; namely a weaker representation of the left side of space within each reference frame.

The latter interpretation received further support from a study by Driver et al. (1994). They used stimulus patterns and a paradigm introduced by Palmer (1989) to study the influence of global and local reference frames on orientation perception of the whole display. Equilateral triangles were placed in various orientations on a page and were grouped by either their bases or their axes (Figure 2.7). When grouped by their bases, the triangles appear to point in a direction perpendicular to the bases, but when grouped by their axes they appear to point in a direction along the alignment of the axes. Palmer (1989) suggested that these effects were due to the way in which global and local properties interacted to produce an overall frame of reference. If this is so, then information to the left of the direction to which the triangles appear to point should more likely be neglected in patients with right hemisphere damage and left neglect than information to the right. This was indeed the case. Detection of a small gap in one of the triangles was more often missed when it was to the left of the direction to which the triangles appeared to point than when it was to the left in viewer- or page-centered coordinates.

A remarkable example of putatively object-based neglect was described by Halligan and Marshall (1997) in a patient who was an accomplished artist (Figure 4.18a shows an example of one of his sculptures before his stroke). After his stroke he produced the example in Figure 4.18b. Not only did he leave out a large part of one side of the face and head that he was molding, he did so even though the clay was on a turn wheel and despite the fact that he worked on the bust from different vantage points. But notice that in this case, the coordinates are not strictly object centered.

A

B

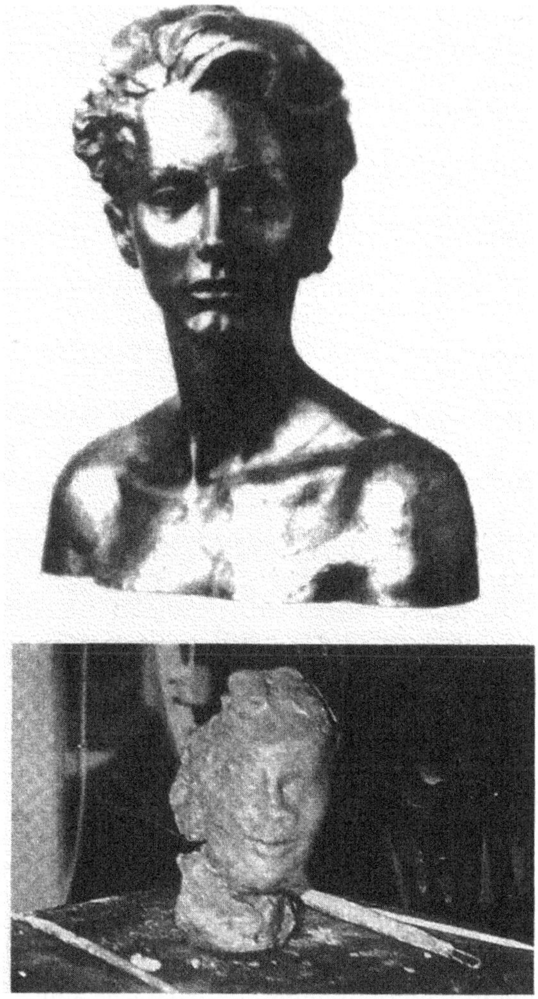

FIGURE 4.18. Sculptures by an accomplished artist before (a) and after (b) a stroke resulting in left neglect. (From Halligan, P.W., and Marshall, J.C., The art of visual neglect. *The Lancet,* 350, 139–140. Copyright © 1997 Elsevier Science. Reprinted with permission.)

If they were, the left side of the sculpture itself should be missing rather than the left side as seen from the perspective of a viewer looking at the face head on. The left was defined in relation to the left as seen from the artist's mind's eye. Nevertheless, the object's left side from this perspective remained the left side independent of the various spatial transformations that occurred during the sculpting process.

Many other studies of object-based neglect have been reported (see Behrmann & Haimson, 1999, for a review), and in the majority of cases, a multiple frames interpretation can explain the data as well as, if not better than, an object- versus space-based attentional account. Nevertheless, there is at least one exception that is problematic. In a case reported by Halligan and Marshall (1993b), a patient with a large right middle cerebral infarct was noted on clinical evaluation to have left neglect. When

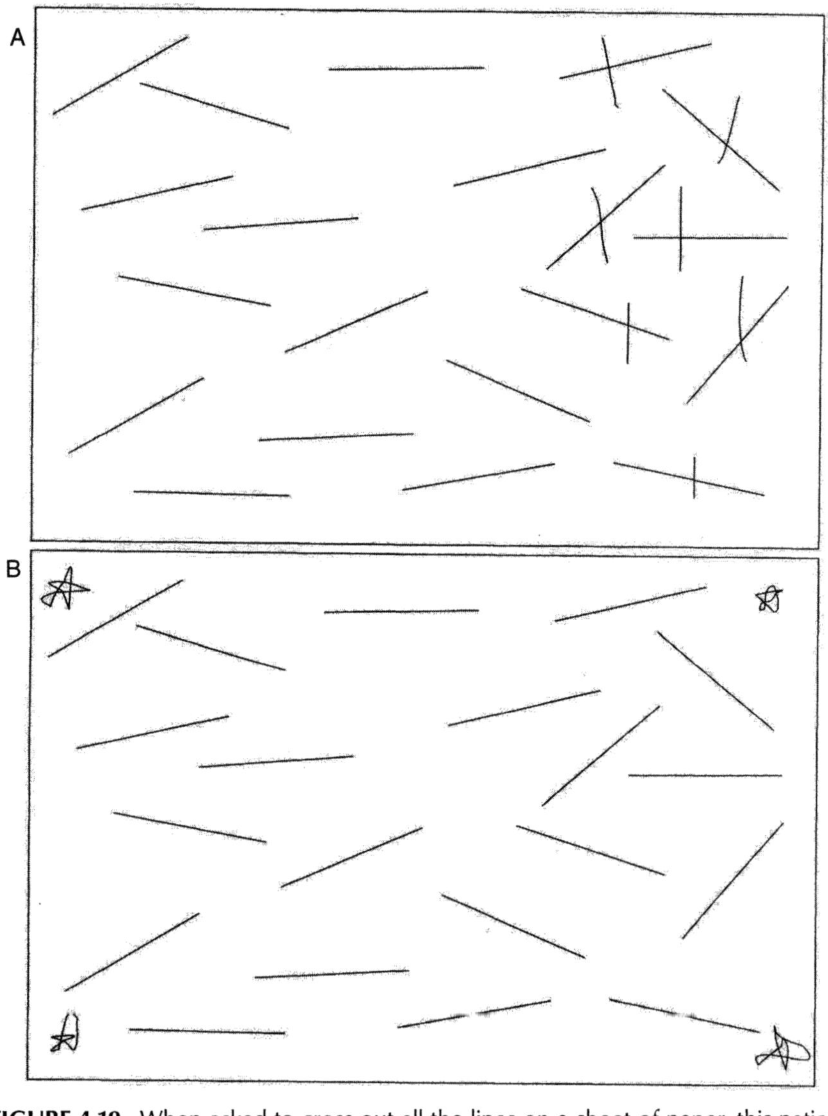

FIGURE 4.19. When asked to cross out all the lines on a sheet of paper, this patient with left neglect only crossed out seven of the rightmost lines (a), but when asked to mark the corners, the patient was able to do so successfully (b). (Adapted from Halligan & Marshall, 1993.)

given a standard bedside test for neglect that requires crossing as many lines as possible on a sheet of randomly oriented lines (Figure 4.19a), the patient showed the typical pattern of left neglect. He crossed out lines on the right side but missed lines on the left. However, when given a new sheet of paper with the same lines and asked to mark only the 4 corner lines, the patient was able to do so (Figure 4.19b). Nevertheless, as soon as he was asked again to cross out as many lines as possible, he reverted to missing lines on the left side. It was as if he could not maintain global attention.

If neglect were a spatial selection problem in different frames of reference, one would expect that the left two corners of the page of lines would be missed in the four-corner condition (global). It appears that this patient could selectively attend to the global configuration when asked to do so and that the spatial referents within this frame were intact, but when asked to attend to the local elements, the left side of space seemed to disappear. Halligan and Marshall (1993) suggested that the problem was one of seeing the whole when attention was constricted and focused on local elements (in this case individual lines). They concluded that the patient could "see the whole array but only a lateralized subportion of the 'objects' that make up the array." Although Halligan and Marshall did not conclude that there was an object- and space-based mode of attention, their data have been used to argue for lateralized attentional differences of attending to objects and space (consistent with Egly, Driver, & Rafal, 1994; Egly, Rafal et al., 1994).

Nevertheless, there is another puzzling aspect to this case. When asked to match the stimulus display size (that incorporated all the lines), the patient chose a display size that was half the size of the display itself. In other words, he chose a size that conformed to an area over which he crossed out lines when locally directed. It was as if the display as a whole had been cut in half and he neglected the left side. His ability to mark the four corners remains a mystery, but it might be that having been asked to pay attention to corners, he used a strategy of tracing the edges of the paper itself until a corner appeared or switched to a more global frame such as the screen or the room as a whole. In the first case, the line closest to the corner he visually traced would then be crossed. This would be a piecemeal type of account, something like that reported by Humphreys and Riddoch (1987) in a case of integrative agnosia. Another possibility is that the patient switched to a global frame, similar to the account given by Halligan and Marshall, but in this case the frame may be as large as the room itself. Admittedly, these are poor attempts to explain the phenomenon, and it would be helpful if other cases of this sort replicated these results, and the anatomy and etiology were better known.

A

B

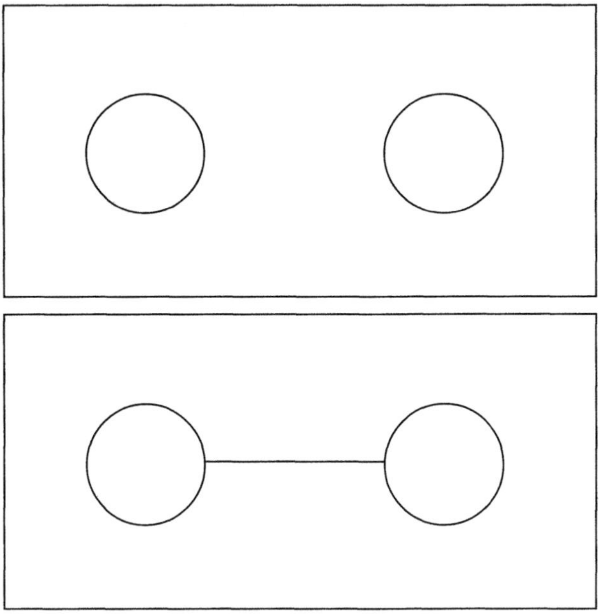

FIGURE 4.20. Neglect of the left circle is more likely in (a) than in (b).

It would also be amiss not to mention the many examples showing that connecting elements presented on the left and right side of a display can change the magnitude of neglect (Behrmann & Tipper, 1999; Driver, Baylis, & Rafal, 1992; Gilchrist, Humphreys, & Riddoch, 1996; Mattingley, David, & Driver, 1997; Ward, Goodrich, & Driver, 1994). For instance, connecting the two circles in Figure 4.20a by a line that creates a barbell (Figure 4.20b) can reduce how much of the left side is neglected. Even illusory contours can reduce neglect of the left side (Mattingley et al., 1997).

Is this due to attention now being directed to objects or can it be explained in terms of the object/space hierarchy discussed throughout this book? Consider the frames of reference that would define spaces within the display of Figure 4.21a. The computer screen and the two circles define the hierarchy. Placing a spatial coordinate on each of the objects in the field (and assuming that the origins have been shifted to the right as expected with left neglect) would result in something like Figure 4.21b. In Figure 4.21c, the circles have been connected to form a barbell, creating a new and third object/space to be described. The additional frame centered on the barbell with its own shift to the right would result in a space/object hierarchy that is different from when the circles were presented alone (Figure 4.21d). If all three frames worked together to influ-

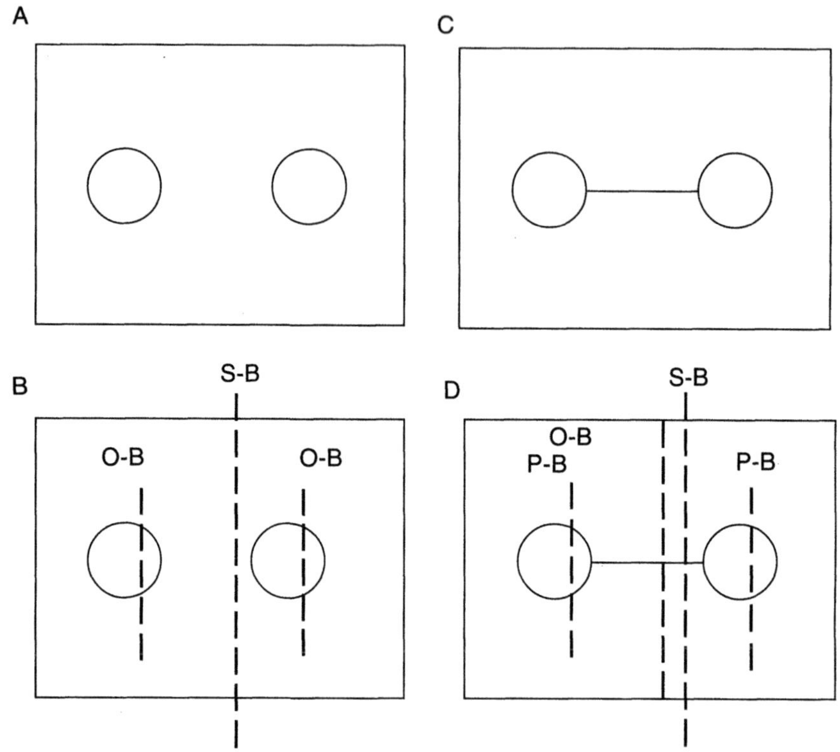

FIGURE 4.21. Example of how multiple frames could interact to reduce neglect (see text for details). O-B, object-based; P-B, part-based; S-B, screen-based.

ence attention, then one would expect less neglect in Figure 4.21c than in Figure 4.21a, if for no other reason than that the additional spatial frame produced by the barbell would pull the center of the display to the left compared to Figure 4.21b. Averaging over the biased vertical displacement of the frames would produce a reduction in neglect in the overall display.

Although this explanation of object-based effects is admittedly post hoc, there is other evidence consistent with it. Changes in aspect ratio of the two sides of a horizontal line changed the magnitude of absolute measures of left neglect (Chatterjee, Mennemeier, & Heilman, 1994; Marshall & Halligan, 1990). Patients with left neglect typically bisect such lines at a point to the right of true center when asked to place a vertical mark at the exact center of the line, something like that shown in Figure 4.22a. Importantly, they do not bisect lines placed at different places on a page, like that shown in Figure 4.22b, which would indicate a fairly stable dominance of the right side of a unitary space. Rather, they are more likely to bisect the lines something like that shown in Figure 4.22c. This finding

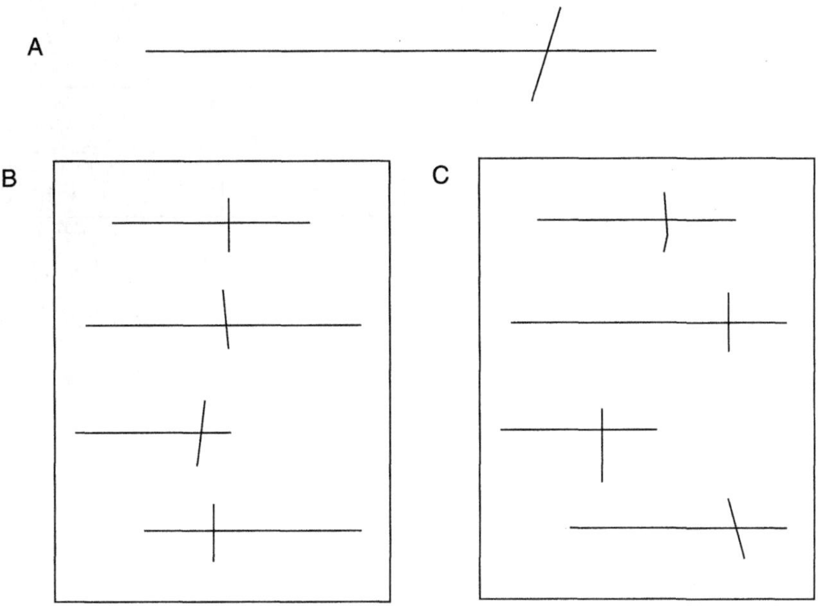

FIGURE 4.22. Patients with left neglect typically bisect a horizontal line to the right when asked to bisect the line directly in the center (a). When lines are placed in different positions on a page, they are not likely to cross the lines as shown in (b), which would be indicative of viewer-centered neglect. Rather, their performance is more like that shown in (c).

suggests that in neglect the shifted origin of the spatial center is proportional to the line and is centered on the line, not the viewer.

Nevertheless, the aspect ratio of the line length itself is not sufficient to account for all line bisection data in patients with neglect. A patient with neglect who showed systematic line bisection as an aspect ratio with longer lines violated the aspect ratio rule with shorter lines (Halligan & Marshall, 1998). In this study lines of different lengths were presented to the patient in the center of a display and the patient was asked to mark the center of the line in each display as usual. The aspect ratio for line crossing did not remain constant for all line lengths. But note that the aspect ratio can account very well for all but the two smallest lines (Figure 4.23).

The reversal with short lines has had a large impact on the neglect literature. However, there is a potential explanation in terms of spatial neglect in global and local reference frames that might explain these effects. When a line is placed in front of a patient, it is on a background, or a more global structure such as a sheet of paper or a computer screen. The dimensions of the global background typically remain constant throughout a testing session, so that a line that is 7 inches long may look something

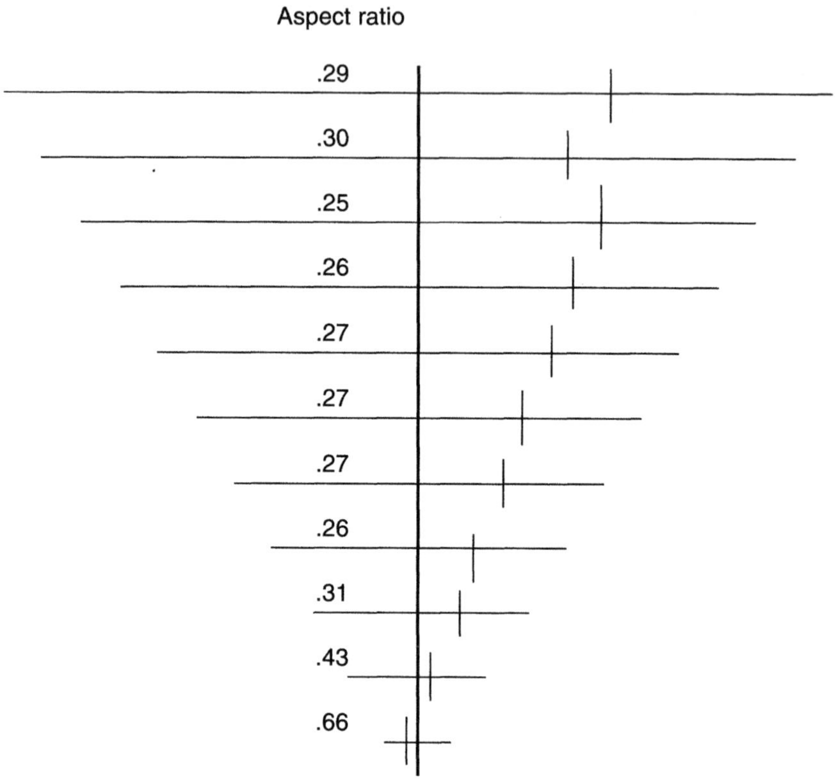

FIGURE 4.23. A study of line bisection in patients with neglect demonstrated a systematic reduction in the proportion of the line that was neglected as line length decreased. (Adapted from Halligan & Marshall, 1988.)

like Figure 4.24a, almost touching the edge of the page, while a line that is 2 inches long may look like Figure 4.24b. The background display defines one spatial frame, while the line itself defines another. If the combination of the global and local frames, like that shown in Figure 4.21b and 4.21d is taken into account, then at short line lengths the average of these frames will fall in an area of the page that has no portion of the line at all. The patient would either miss the shorter line entirely or initiate a search for it, possibly producing an overcompensation or overshoot of the midline. If this is the case, one would expect more eye movements to the left with short lines than long lines, and this is something that would be quite easy to test.

Although this reason for the overcompensation is speculative, the combination of global and local frames can quite nicely account for the increase in the aspect ratio with short lines. Whatever the explanation of

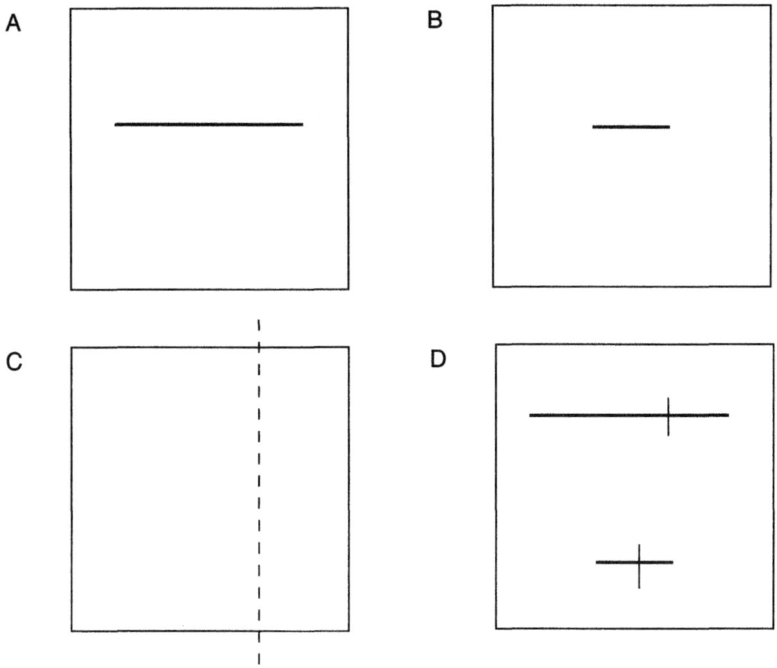

FIGURE 4.24. Results like those represented in Figure 4.23 might be due to a constant aspect ratio of neglect as shown by others, but one that combines global and local frames of reference (see text for details).

the short line effects turns out to be, the data show that aspect ratio as an important feature in neglect performance, at least with longer line lengths.

What Is an Object for Object-Based Attention?

In chapter 3, I discussed some of the different ways in which investigators studying spatial attention have conceived of the space attention uses for selection. Some investigators seem to think of this space as everything that is not an object for perception at any given moment. Others reserve it for the most global parts in a display that are outside whatever could be potentially perceived as an object. Others tie space to the retina, early cortical maps, or receptive fields, but only a few have asked whether space has multiple layers or multiple maps within the brain (although see Graziano & Gross, 1994, 1998, and Gross & Graziano, 1995, for a neurobiological theory based on multiple spatial maps that are somewhat different than the object/space proposed here, and also Mesulam, 1999, for a discussion of multiple spatial maps in producing hemineglect).

Objects are as difficult as space to objectively define, but the idea of multiple objects existing at multiple levels has long been acknowledged. We seem to have an intuitive idea of what an object is, namely a whole or a unit that maintains its coherence even when it moves or we move. This description does not leave much, if anything, that is not an object, although there are perceptual units we would not typically call objects (e.g., holes in the ground, slits in a sheet of paper, a pen mark on the sofa). Our intuitive notion of what an object is generally does not include the view of the San Francisco Bay from the Berkeley hills or the forest through which we are hiking. Is an object anything we call an object, or is it something else?

There has been a long history and volumes of papers written on the topic of object perception and recognition, and it is not my intention here to discuss various theories of perceptual organization and the nature of objects. Rather, my goal is simply to raise the question of what we mean by an object for attention. For instance, when we talk about object-based attention associated with ventral processing streams in the cortex, what exactly is being selected and to what extent does object selection (as opposed to perceptual organization or feature extraction) depend on the integrity of the temporal versus parietal cortex? How can we know when attentional selection is object-based without knowing what an object is?

This is not to say that we need solid definitions of objects and space before experiments proceed that address different modes of attention we might conveniently call object- and space-based, but it should raise a flag concerning being too glib about what we consider as a good object with which to test object-based theories of attention. It should also stimulate questions about what "objects" emerge in awareness when spatial attention is dysfunctional. To what extent does this object/space hierarchy I have been discussing rely on the ability to spatially select objects, or spatial frames?

In the tradition of the Gestalt psychologists, objects in cognitive science have been thought of as closed, connected, or grouped features that form perceptual units either separate from a background or separate from other perceptual units (Koffka, 1935, see also Robertson, 1986, for a more contemporary view). For instance, in Figure 4.25, the perceptual units might at first look like white blobs scattered over a black background, but further inspection reveals that perceptually filling in the edges between the blobs produces a new perceptual unit (in this case a face). We find that the space between the original perceptual units is not a blank field after all. It contains clues to spatial structure, and not only do the number of perceptual units change from many to one when we see the face, but the black background changes from one to two spaces (the space outside and inside the face). What once was a unitary space is now part of two spaces.

FIGURE 4.25. Do you see a face in this figure? Example of perceptually filling in by closure between gaps.

The object of attention has changed as a result of attending to the space between the blobs. Does this start to sound circular? If it does not, then I have failed.

This circularity has a way of creeping into theoretical claims about modes of attention. One example should suffice, and I will refer to a paradigm that has played a prominent role in this chapter. Patients with commissurotomy (split-brains) produce no Egly object-based effects (Figure 4.10) in a cuing study when a pair of rectangles is presented so that they are projected directly to the right hemisphere, but they do show the normal effects when presented to the left hemisphere (Egly, Rafal, et al., 1994). These findings have been used to support other evidence from patients with right or left hemisphere damage. The conclusion was that the left hemisphere attends to objects while the right attends to space. When combined with evidence for dorsal/ventral differences in space and object processing, the question then arose as to whether the dorsal stream is dominant in the right and the ventral dominant in the left hemisphere.

However, it could be that the bias for global information in the right hemisphere would create a slightly different initial spatial representation of the stimulus than the bias for local information in the left hemisphere (see Ivry & Robertson, 1998). In this case the difference in results in the split brain would not reflect selection of space on the one hand and objects on the other, but instead reflect differences in the spatial bias for global structure (i.e., the square formed by the two Egly rectangles) in the right hemisphere and for the local space (i.e., rectangles) in the left. Differences in the spatial organization of the stimulus would produce differences in how attention would be allocated within globally and locally defined spatial frames of reference. Attending to space that defines a global square configuration produced performance that was interpreted as space-based attention, while attending to space that defines a local space produced performance that was interpreted as object-based. Within the scenario I have been discussing throughout this book, the differences between the hemispheres would not be due to object- versus space-based attention, but to the perceptual organization of the stimulus within a hierarchical space/object framework.

5

CHAPTER

Space and Awareness

Without an intact sense of space—whether speaking of the space that defines an object or the space that defines a whole scene—how would the world appear? Could we be aware of an object at all? After all, objects do not exist without a location in the outside world, but, as it turns out, they can in the internal world of our minds.

Issues concerning the relationship between space and awareness have generated a great deal of interest in cognitive neurosciences, largely from the observation that brain damage can produce striking spatial deficits that affect awareness itself. These deficits are most obvious in neuropsychological syndromes such as unilateral neglect and Balints syndrome. The very existence of unilateral neglect demonstrates that awareness of parts of space can disappear from perception at the same time as other parts remain intact even when there is no primary sensory loss (see Driver & Vuilleumier, 2001, for an excellent overview of the cognitive and neurological mechanisms of awareness and spatial neglect).

The loss of awareness of one side of space but not the other in the same person at the same time is an extraordinary phenomenon and one that is nearly unimaginable for those of us who have not experienced it. In fact, it is rather unimaginable for the patients themselves too. When asked about their deficits, patients who suffer from hemineglect often say something like, "They tell me I am not attending to the left " rather than "I can't make out what is on the left side." It is as if they accept that there is a left only because someone else like the professional staff and/or families insist on it. The patients themselves can appear phenomenally unaware of left items. Consistently, they often show surprise when told that a picture they have just drawn was not finished on the left side. Their explicit perceptual worlds are defined by the spatial maps that their remaining neural apparatus is able to bring into conscious awareness, and these do not include large portions of the contralesional side of space.

Not attending to a part of space in normal individuals does not have this quality of space disappearing from conscious awareness. While you are reading this sentence, you may be ignoring many words on the page and most of the information beyond this page, but you know you can move attention to those places whenever you choose. You realize that if you turn your head and body 180°, information that is currently behind you will come into view. But for patients with unilateral neglect, the spatial world has narrowed or changed in ways that can eliminate parts of it from this type of awareness. In this regard, neglect appears to be much more than an attentional deficit. Clearly attention is affected because people do not try to attend to places they don't believe exist, but there is a qualitative difference between not attending in cases of neglect and not attending in a normal perceiver. There can be disbelief or surprise by patients with neglect that they do indeed miss one side of a display.

Despite the conscious loss of portions of space, there are many experiments that have shown that patients with neglect can process information on the neglected side below the level of awareness. In vision these include properties such as figure–ground organization (Driver, Baylis, & Rafal, 1992), grouping (Ward et al., 1994), color and shape similarity (Baylis, Rafal, & Driver 1993), perceptual illusions (Mattingley, Bradshaw, & Bradshaw, 1995), and even semantic information (Berti & Rizzolatti, 1992; Ladavas, Paladini, & Cubelli, 1993, McGlinchey-Berroth, Milberg, Verfaillie, Alexander, & Kilduff, 1993). For instance, both McGlinchey-Berroth et al. (1993) and D'Esposito, McGlinchey-Berroth, Alexander, Verfaillie, & Milberg (1993) showed that words that were neglected on one trial affected reaction time on a subsequent trial. When the two words were semantically related, response times were faster than when they were unrelated, despite the fact that the first word was neglected and the second was not.

Of course all stimuli, including words, have shape that can be defined by the 2-D spatial topography of the retina. Each word contains letters appearing in a string, and each letter itself can be described spatially. So it seems that the spatial locations of letters in words, as well as the spatial structure of each letter, must be coded in some spatial representation whether or not the semantics of the word are implicitly or explicitly registered. The implicit encoding effects found with neglect support a view that space itself is left intact, while other mechanisms such as those that support spatial attention are not. Given the range of stimuli that produce implicit effects in the neglected field, it is enticing to conclude that the reason information on one side of space (say, the left) is not explicitly reported is because these patients do not move attention to the left side while attention continues to move to the right. Under conditions when patients do move attention into left space, then the stimulus is no longer neglected and patients sometimes express surprise that they hadn't seen it before.

But why are they surprised? Actually, it could be argued that an analogous situation exists in normal perceivers under the right conditions. For instance, the surprise that is exhibited when changed events in the world go unnoticed (called "change blindness") could be likened to that shown by patients with neglect. In normal perceivers, the disappearance of even very large or obvious items in a scene can be completely missed (Rensink, O'Regan, & Clark, 1997), and subjects are often astonished when they discover that they did not see, for instance, a large engine disappear from the photo of a DC-10. It is enticing to relate change blindness in normal individuals to neglect in terms of spatial unawareness, and especially so because recent imaging evidence suggests that spatial attention and dorsal, "where" functions are involved in change blindness in normal individuals.

Beck, Rees, Frich, and Lavie (2001) used functional imaging (fMRI) in a change blindness paradigm and found that the parietal and dorsolateral prefrontal cortex (the same areas most often implicated in hemineglect) are activated when changes in a display are detected by normal perceivers but not when they are missed. Behaviorally, increased attentional demands at one spatial location increased change blindness at another. When attention was directed to the location where change occurred, events at that location were noticed, and when attention was directed elsewhere they were not.

Most importantly for the topic of this chapter was that Beck et al. (2001) also found ventral activation in the temporal lobes independent of change detection (i.e., below the level of awareness). Temporal activation alone was not sufficient to perceive the change, although temporal activation was evident whether or not the change was detected. In contrast, an increase in parietal activation was only observed when changes were detected.

How might this information be applied to hemineglect? If parietal lobe damage reduced the attentional resources available to attend to the whole of space, it would bias attention to one side at the expense of the other. In the case of right hemisphere damage, attention would be biased toward the right side, neglecting the left. Some have argued that spatial representations per se are not affected by parietal damage. Rather, attentional mechanisms are directly affected, producing what then is manifested as a spatial deficit on clinical tests. The logic of this argument is very compelling on the surface, especially considering the strong evidence for implicit visual and semantic information in the neglected field. Space must be sufficiently represented in order to define the shapes that are implicitly encoded. But notice that underlying this argument is the assumption that space is a unitary whole, and we have already seen the fallacy of this assumption in previous chapters.

If the evidence for multiple spatial maps is considered, the question of attention versus spatial representation deficits as the underlying source of hemineglect again becomes an issue. Is it possible that some spatial maps are represented below the level of awareness, while others are not? Are there implicit spatial maps that might support the many and variable implicit effects that have been observed in patients with unilateral neglect? In other words, are there spaces that remain intact independent of attention? One way to test the hypothesis of implicit spatial representations would be to test an individual who is not explicitly aware of the location of the objects he or she perceives and attends to. As I've already discussed in many of the previous chapters, we have been studying such a patient (RM) for some time, and have in fact found evidence for intact implicit space in the face of severe explicit spatial deficits.

☐ Spatial Functions of a Balints Patient

Recall that the neurological patient RM was diagnosed with Balints syndrome after strokes that affected both parietal lobes (Figure 5.1). He was 54 years old when he suffered his first stroke in the right hemisphere, and some months later he had a second stroke that affected his left hemisphere. Both strokes were most likely embolic in nature (blood clots). The emboli obstructed similar branches of the middle cerebral artery in each hemisphere, producing occipital-parietal lesions that were very nearly symmetrical. The location and extent of damage were similar to that shown in Figure 1.5 and reported by Balint (1909), who was the first to note the link between bilateral parietal damage and the behavioral profile that defines Balints syndrome (see chapter 1).

The most striking observation in these cases is the nearly complete loss of spatial information that nevertheless does not affect the ability to report the identity of a single object (simultanagnosia). This profile has been consistently reported across the relatively few patients with this syndrome who have been studied since Balint's original report (see De Renzi, 1982; Rafal, 1997; Rizzo & Vecera, 2002, for reviews).

Over the course of time we documented RM's explicit spatial deficits in various ways. When we began testing him in 1992, it had been several

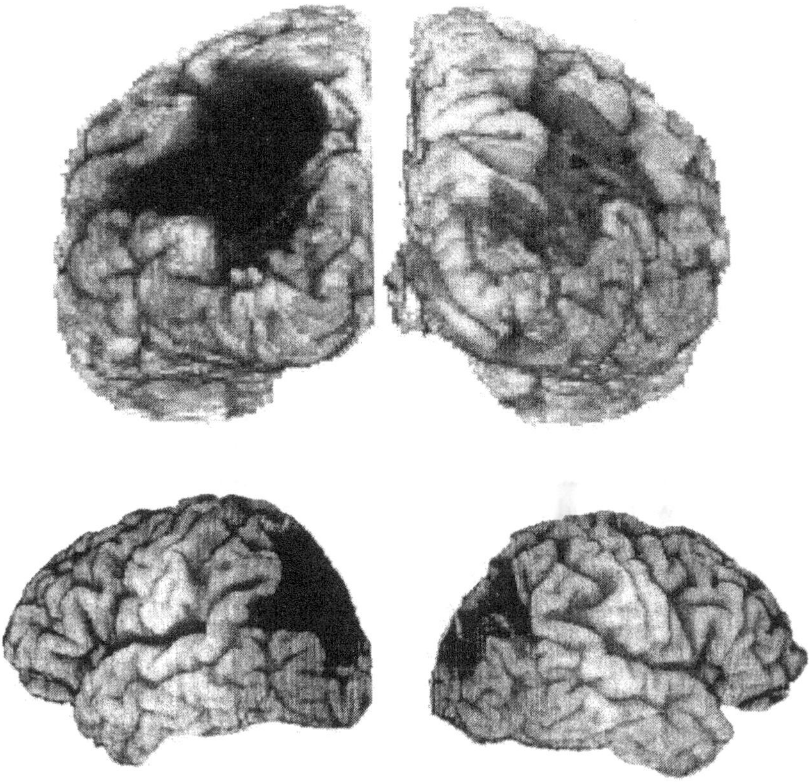

FIGURE 5.1. Reconstruction of RM's MRI images showing bilateral parietal-occipital damage. Note that the supramarginal gyrus of both hemispheres has been spared.

months since his second stroke, and he was neurologically stable but with Balints symptoms that were severe and classic. During the first testing sessions he was unable to accurately locate anything he saw better then chance even under free viewing conditions, either verbally or by reaching for an object or by pointing in its direction. He was not able to move about without guidance. It was as if he were completely blind, but in some sense worse, for he was unable to code spatial relations between the outside world and himself or between one location and another. Although his auditory spatial knowledge was not tested initially, a later controlled study in a soundproof environment demonstrated that spatial abilities in localization of sounds were also affected, although less severely than his visual spatial performance (Phan, Schendel, Recanzone, & Robertson, 2000).

Formal visual evaluation showed that RM's visual acuity was 20/15 in both eyes (without glasses). He had normal contrast sensitivity, normal color vision, and full fields (an early perimetery test suggested loss of vision in a crescent of the lower field about 10 degrees from fixation that was absent on subsequent perimetery tests). For all intents and purposes RM's sensory vision was extraordinary for a man of his age. A formal audiogram given at the time of auditory testing showed normal hearing as well. As with other Balints patients, RM exhibited optic apraxia, a condition in which the eyes remain fixated (usually directly straight ahead) in the absence of any ocularmotor deficit. Balint called this symptom "psychic paralysis of gaze" because when he turned the patient's head manually, the eyes moved normally in their sockets while maintaining their fixation at the same location in the external world. This was the case for RM as well, at least during the early days of testing.

Single objects popped in and out of view in RM's everyday life. Consistent with reports in the literature, an object continued to be perceptually present for a while and then was replaced by another object or part of an object without warning. However, the spatial location of the object or part he perceived at any given moment was unknown to him. RM was unable to accurately reach in the direction of the object he saw (whether with his right or left hand), producing random movements until his hand happened to bump into the object (optic ataxia). He would then readily grasp it. Neither could he verbally report the object as being to the left or right of him or towards his head or feet. His location errors were not due to spatial confusion, as he could readily report that his right or left hand or the right or left or upper or lower part of his back had been touched. He would accurately follow instructions to touch his upper left arm with his right index finger or to grab his right ear with his left hand. He could also follow commands to move his eyes or hands to the right or left, up or down, although eye movements were initiated slowly. The spatial frames of his body were intact. Despite an intact body-centered frame of reference, he was dismal at determining where items were that were placed in front of him even when they remained in full view. This problem could not be attributed to confusion or comprehension difficulties, as RM's language, memory, and judgment were within normal limits. He was capable of making his own decisions and engaging in conversation; he remembered details of each testing session, where and when they happened; and he was able to recognize and recall the names and faces of the many students and colleagues who had examined him, sometimes when they returned 2 to 3 years later.

During early testing of his extrapersonal spatial abilities he often made statements like, "See, that's my problem. I can't see where it is." He also found it hard to describe what his perception was like. His explanations suggested that objects that popped into his view were not mislocated per se. Rather, they simply had no location in his perceptual experience. Despite these explicit spatial problems, we found several indications that his brain encoded where things were even though he was not explicitly aware of where they were (i.e., he showed evidence of implicit extrapersonal spatial representations). The evidence supporting intact implicit spatial encoding will be described later.

RM's explicit spatial problems were most severe when we began testing him (a few months after his second stroke), and some spatial functions slowly recovered over time, but by no means have they ever been close to normal. Because his spatial problems were evident in his everyday life he required constant care and guidance until his spatial abilities recovered sufficiently for him to manage some of his basic needs by himself (see Robertson et al., 1997, for full history). Cases such as RM's form the major support in the literature for the existence of implicit spatial maps in the face of severe explicit spatial problems produced by parietal damage, but these will be discussed later.

☐ Explicit Spatial Maps

Explicit Access to Space with Unilateral Damage

Balints syndrome is rare, at least in a form in which it can be carefully studied. It can be observed in some dementing diseases such as Alzheimer's, but other parts of the brain are compromised as well, producing memory and other deficits, therefore making it difficult to study systematically. Thus, the data collected to date with patients with Balints syndrome are limited, and most of the evidence concerning the effects of spatial awareness after brain damage have been derived from studies of hemineglect. In severe cases of neglect, nothing is reported as being present on the contralesional side of space. It is well documented that even on occasions when an item is seen on the neglected side, its location may remain uncertain. For example, in an early study Hannay, Varney, and Benton (1976) briefly presented dots in the left or right visual field followed 2 seconds later by a display of numbers, and patients were asked to read the numbers that were at the locations of the dots. Patients with right hemisphere damage made significantly more location errors than either patients in another brain damaged group or healthy controls. In a more recent study reported by Vuilleumier and Rafal (1999), patients with right hemisphere damage reported the number of items in a 1- to 4-item display equally well whether items were presented on the left or right side or both, while locating the items was extremely poor. In other words, even when subjects detected an item, its location was not always known.

A clinical sign that is consistent with these effects is known as *allesthesia* (perceiving a stimulus presented in one location as in another location), which is prominent in some patients with hemineglect. This phenomenon is quite remarkable to observe. A patient can be very certain that a tap on his left hand was actually on his right or that a visual stimulus shown on his left was presented on his right. A patient might point to a place where nothing appears and say, "Yes, it is right there" even while it remains clearly present on the left. Dramatic spatial transformations such as these are consistent with arguments that the underlying problem in hemineglect is a spatial one and only secondarily an attentional one (Bisiach, 1993; Bisiach & Luzzatti, 1978).

Spatial deficits in hemineglect have also been investigated in audition, and a recent study reported striking dissociations between detection of sounds and their locations. Both behavioral and electrophysiological measures with hemineglect patients demonstrated that when the task was to detect pitch or duration of a sound, performance was equal whether the sound was on the right or left side. Components of the evoked related potential (ERP) specific to pitch and duration were normal over both hemispheres as well. However, locating the sound was better when it came from the right than the left side (Deouell, Bentin, & Soroker, 2000). Consistently, abnormal evoked potentials were evident when location was to be reported and the sound was on the left. Similarly, in the visual domain simple features pop out from a crowd of distractors in visual displays even in the neglected field (Brooks, Wong, & Robertson, 2000; Esterman, McGlinchey-Berroth, & Milberg, 2000) but localizing the features is more difficult (Eglin et al., 1989).

Unilateral Versus Bilateral Lesions

There have been some elegant computational models describing how unilateral damage to spatial representations could affect a wide range of processing mechanisms, including those that produce both space- and object-based neglect. For instance, Driver and Pouget (2000) suggested that a gradient across the visual field, something like that shown in Figure 5.2, could be produced by damage to the right hemisphere. If an object were shown anywhere along this gradient, the strength of response on the left side of the object would always be stronger than on the right side of the object because the *relative* strength would differ across the object at any two points. Although this model may account for the per-

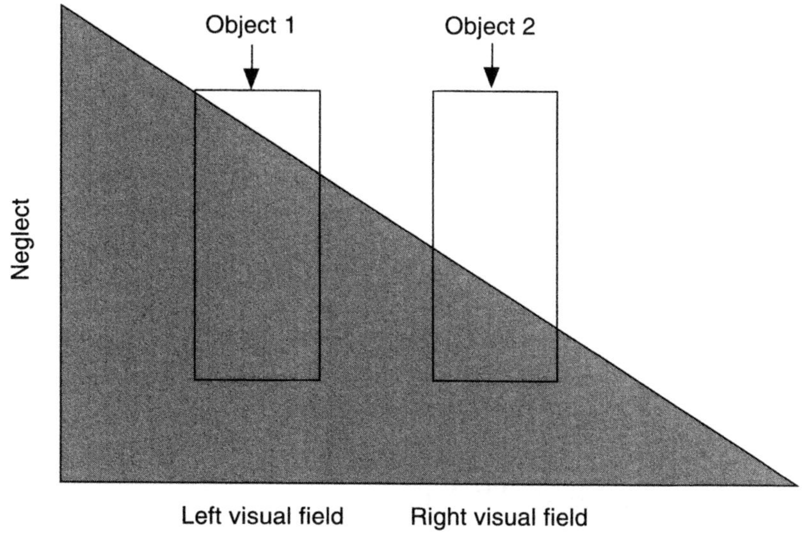

FIGURE 5.2. Schematic of how neglect across the left and right visual fields could produce object-based neglect. An object appearing at any point along the diagonal line would result in more left-sided than right-sided neglect. (Adapted from Driver & Pouget, 2000.)

formance of patients with neglect on some tasks, it does not account for the occurrence of neglect in frames rotated out of alignment with the viewer or for allesthesia. It also is unclear how models of this type would account for the way in which patients with bilateral parietal damage and Balints syndrome perceive the world. The object a Balints patient sees is sometimes presented at fixation but sometimes in the periphery. For instance, RM often would report that he was looking directly at an object he saw even when it was presented off to his left or right side and his eyes were fixated straight ahead. Also, he was likely to report an object's location as "central" under conditions where he was forced to guess.

Another difference between bilateral and unilateral parietal damage is that a patient with Balints syndrome, such as RM, is aware that there are objects in space that he cannot see, while patients with neglect often miss items on their left side and act like that side of space has disappeared. Many investigators have argued that unilateral neglect is a less severe version of Balints syndrome (see Farah, 1990). RM's deficits were not of this sort and were closer to that of unilateral extinction in this regard. For instance, patients with left extinction perceive single stimuli presented to the right or left equally well but miss left items when stimuli are presented on the left and right sides together. The item on the right side seems to capture their attention at the expense of the item on the left.

These patients remain aware that a left side exists but miss items on the left when right-sided stimuli are presented at the same time. Qualitatively, they are often unsure of whether they saw something on the left side or not. Similarly, RM did not report any objects outside the one that entered his awareness, but he was aware that a space existed outside the one object he saw. Other objects in the field seem to be extinguished by the object that is attended. Unlike neglect patients, RM could not move his attention to other objects voluntarily. If that object disappeared (either because the experimenter or RM's visual system removed it from view), another would take its place. RM did recover limited spatial abilities over the years we tested him. Whether his partial recovery was due to the concentrated testing, his own persistence and creativity, or time itself is unknown.

Explicit Access to Spatial Relations

In systematic tests of RM's spatial abilities after some initial spatial recovery, we found that he was about 70% accurate in reporting whether a single X that remained on a computer screen until he responded was toward his right, left, head, or feet or centered on the screen in front of him. The X was presented in one of nine different locations along horizontal and vertical axes and stayed on the screen until RM responded. In another condition when we asked him to report whether an X and an O were in the same or different locations when presented sequentially, he was at chance even when he clearly saw both. When the X and O appeared simultaneously and he judged whether the X was to the right or left, above or below the O, he was at chance. Again, the two letters stayed on the screen for as long as RM needed. His above-chance performance in locating a single X on the screen seemed to reflect some recovery in relating an X either to his body or to the more global environmental cues, while his ability to relate one object to another within a single frame (even on the trials when he saw them both) was severely impaired (Friedman-Hill et al, 1995; Robertson et al, 1997).

These findings suggest that the computation of relative locations may be different depending on the spatial frames selected. RM could relate one object to himself above chance levels (the single object condition), so why didn't he use his own body as a referent when judging the same object when in the context of a second? If judging relative location between objects used the same referent as judging their location relative to his body, RM's better-than-chance performance in locating a single shape would predict better-than-chance performance in the relative judgment task, at least when he saw both. This would be expected because he could have first related the X to himself as he did in the single X task, and when he saw the O he could then relate the O to himself in the same way and then deduce the location of the X with reference to the O in his own body frame. There was no evidence to suggest he could do this.

It was also possible that RM learned to see the spatial relationship between the global frame of the screen and a single object on the screen (the single X condition) but not the relationship between 3 items that would include the screen and the X and O on it. So in another study, we asked him to judge whether a black dot was outside or inside a circle. He was unable to do this any better than chance. Even when the dot touched the circle, he was no better than chance. Furthermore, the distance between the circle's perimeter and the dot made no difference at all. Thus, it is unlikely that his ability to locate a single X in the previous study at above-chance levels was due to some recovery of spatial relationships between global and local frames of reference.

Coslett and Saffran (1991) reported similar results in a Balints patient they tested. They asked her to judge which of two circles contained a dot, and she too was at chance in performing this task. In another condition they asked her to report whether a dot in a single circle was in the center or not. Again, she was at chance. Cotlett and Saffron also reported some minimal improvement in spatial skills over time, but like us, continued to observe deficits in judging spatial relationships between a global frame and local elements as well as between two elements in separate locations even after some recovery and under conditions when at least two objects could be seen. As with RM, perceiving spatial relationships within her own body frame remained intact, but relating an external stimulus to her body was impaired.

Loss of a Body Frame of Reference

It is interesting to compare Balints patients with that of a patient, GW, reported on by Stark, Coslett, and Saffran (1996) who had slightly more dorsal bilateral parietal involvement than RM. In contrast to Balints patients, GW seemed to lose the spatial configuration of her own body. One of the more striking problems she had was in positioning herself with reference to an external frame. For instance, when starting from a sitting position, it took her up to 10 minutes simply to work out the body orientation she needed to lie down with her head on the pillow end of a bed. Yet she could accurately point to the pillow and other features in her environment. She did not suffer from simultanagnosia, and upon formal testing, her spatial attention appeared intact. Unlike Balints patients, she was able to serially search for targets in a display and judge the spatial relationship between objects she saw. Other tests revealed that when she was asked to reach for an item, she was accurate when she could see both the item and her reaching hand but at chance when she was unable to see her hand.

Unlike RM, who retained an intact body space but impaired processing of spatial relationships in his internal representation of external space, for GW spatial relationships between objects were intact. However, spatial relationships between the position of her body and the external world were unavailable unless she looked at her relevant body part. These results were consistent with her neurological examination, which found that she could not judge which direction her arm was moved by the examiner when visual input was unavailable, but could do so when looking at her arm.

The differences between GW and Balints patients' spatial disabilities are consistent with the proposal that environmental space is coded in separate maps from body space. Determining the spatial relationships between two objects in the external world and one object and the viewer does not necessarily utilize the same spatial computations. They also differed in the location of cortical damage. GW's lesions were in the superior parietal lobe, while Balints syndrome is associated with inferior parietal damage (see Snyder, Grieve, Brtchie, & Andersen, 1998 for electrophysiological evidence from monkeys that is consistent with this spatial dissociation).

□ Implicit Access to Space

Although the brain appears to utilize multiple spatial maps both cognitive and neurobiological theories have commonly tested only explicit spatial abilities. The question of spatial maps that can function normally outside awareness has only recently been considered, and again has received support mostly from neuropsychological data collected in Balints syndrome.

Like other Balints patients, RM exhibited severe explicit spatial deficits, yet maintained the ability to accurately perceive a single object. In other words, his visual system was able to compute a set of lines that were spatially related to each other to form an intact single unit. Objectively, his attention was drawn to seemingly random objects in his visual field, but further testing demonstrated some systematic influences. First, he was much more likely to see the local details of a stimulus than a global whole. For instance, if a stimulus such as that shown in Figure 3.20 was presented, he invariably missed the global form but was able to report the identity of a local element. When we first began testing him, he was shown

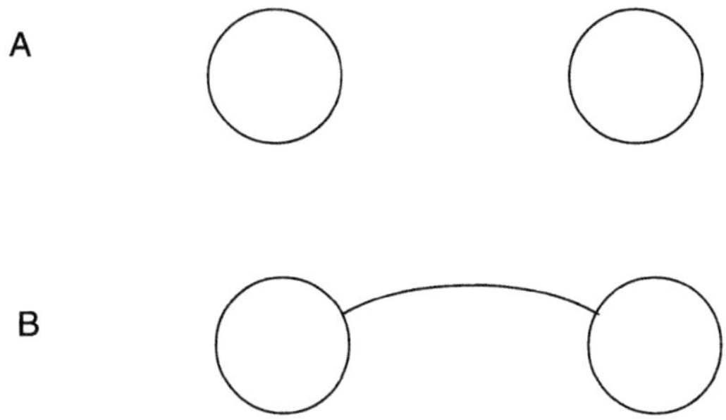

FIGURE 5.3. When two circles were shown to a patient with Balints syndrome, only one was reported (a), but when a line was placed to attach the two circles as in (b), the patient reported seeing "spectacles." (Adapted from Luria, 1959.)

hundreds of these types of stimuli and almost never identified the global form even though the stimuli were in full view and he could examine them for as long as he wanted. He missed the global form even when he was asked to guess or was given a choice between two alternatives. Further testing demonstrated that he was more likely to see shapes that were connected than ones that were not, suggesting that his ability to group unconnected elements into a configuration may have been impaired.

Connecting parts in a display has long been known to change what a person with Balint's syndrome might see (Luria, 1959). For instance, when two separated circles were presented to a Balints patient studied by Luria (Figure 5.3a), the patient reported seeing only one circle, but when a line was drawn as in Figure 5.3b, the patient reporting seeing "spectacles." Humphreys and Riddoch (1993) reported a related effect in a patient with Balints syndrome. When they placed a number of circles on a page, half one color and half another (Figure 5.4a), the patient reported seeing only one color, but when two circles of different colors were connected by a line (Figure 5.4b), the patient reported seeing both colors. Other Gestalt principles of organization have also been shown to affect what a patient with this problem is likely to perceive (Cooper & Humphreys, 2000).

So perhaps it was not surprising that RM saw a local item in a stimulus like that in Figure 3.20 and consistently missed the global shape, since the local items were not connected to form a global shape. We tried several manipulations like these, such as increasing the number of local elements, varying the gaps between local elements, varying the size of the stimulus as a whole, changing the letters to shapes, and so on. We also drew outlines around the stimuli as in Figure 5.5. None of these modifications affected RM's propensity to report the local shape, and he continued to miss the global shape. The fact that density and gap size did not matter seemed to indicate some type of grouping problem with either the

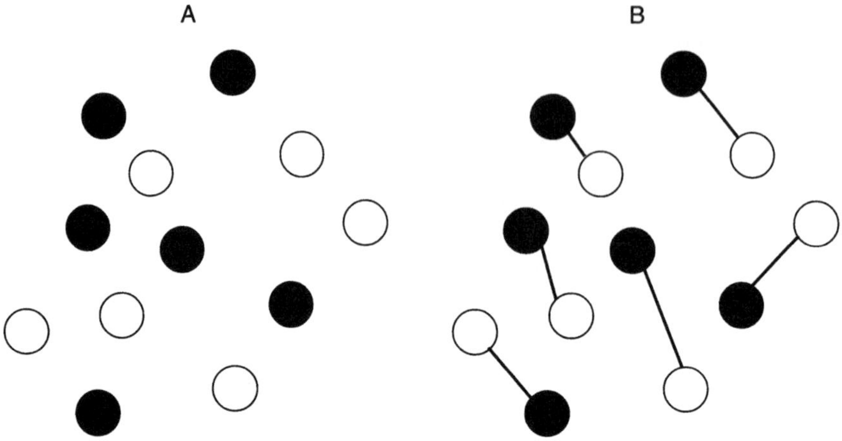

FIGURE 5.4. Examples of displays used to test a Balints patient with each unconnected dot in (a) being connected to a dot of a different color in (b). (Adapted from Humphreys & Riddoch, 1993.)

global form being distorted or the global form simply missed entirely. It turned out that neither of these was quite correct. While explicit access to the correct identity of the global form was severely affected, implicit measures demonstrated that grouping had been achieved.

Implicit Global Processing

In one study we presented hierarchical patterns made from local Hs or Ss to produce a global H or S and measured RM's reaction time to identify

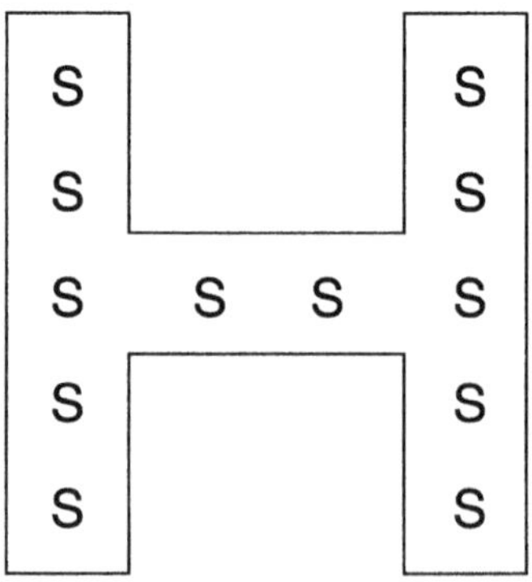

FIGURE 5.5. A Navon shape with global and local levels plus an outline of the global form.

the local form, which he saw readily. This study was based on one first reported by Navon (1977) with normal perceivers in which either the global or local letters were attended in different blocks of trials. Navon's main interest was whether parts or wholes were processed first by normal perceivers, and he found evidence that wholes were processed first. This was supported by two effects. First, global shapes were identified faster than local shapes, and second, global shapes interfered with local response time but not vice versa (Figure 5.6). For present purposes the most important finding concerns the second effect. In blocks of trials when normal perceivers reported the local letters, an inconsistent global shape (e.g., a global S created from local Hs) slowed response time compared to when the two letters were the same (e.g., a global H created from local Hs).

Since RM could report local letters but not global ones, we only asked him to respond to the local letters in a block of trials and measured how the consistency between global and local shapes affected response time (Egly, Robertson, Rafal, & Grabowecky, 1995). Did the global element interfere as it did in normal perceivers? If it did, it would show that the global shape was encoded and by extension that the local elements must have been spatially grouped appropriately at some processing level. The data demonstrated that the global shape was represented below the level of awareness. This finding was confirmed with another Balints patient and reported by Karnath, Ferber, Rorden, and Driver (2000) and has also been observed in a patient with Balints symptoms due to Alzheimer's

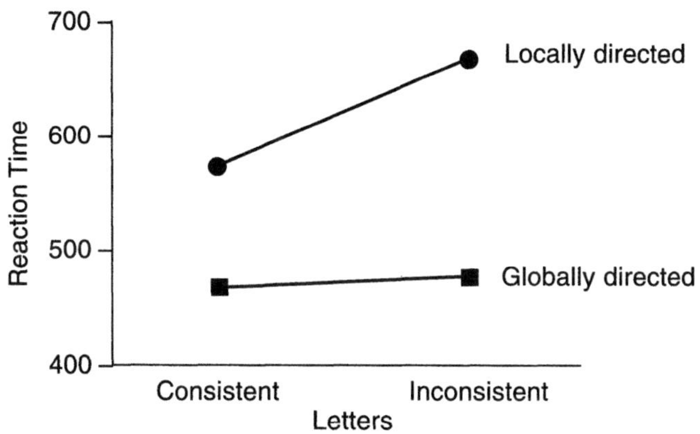

FIGURE 5.6. In Navon's (1977) original study, discriminating global letters was faster than discriminating local letters. In addition, global letters that were inconsistent with the target letter interfered with local responses, but local letters that were inconsistent with the target letter did not affect global responses. It was on the basis of these two effects that Navon proposed his theory of global precedence. (Adapted from Navon, 1977.)

disease (Filoteo, Friedrich, Rabbel, & Stricker, 2002). Even though parietal damage disrupted the ability to explicitly resolve global forms in free viewing conditions, global forms interfered with local responses, as they did in normal perceivers. RM's performance was clearly affected by the shape of the global letter despite the fact that he could not perceive it. Grouping within the global spatial frame was intact (albeit implicitly).

Although this finding was interesting, at first we were puzzled by it. How could a person without a spatial map, who could not even come close to accurately locating a local item he saw, represent a global shape that was dependent on processing the spatial locations of several local elements? The literature in cognitive neuroscience concerning the parietal lobe's role in spatial attention as well as our own studies of RM indicated that RM was not supposed to have complex spatial information, and yet here was the first bit of evidence that he did. There clearly remained substantial spatial processing below the level of spatial awareness in the face of large lesions to the dorsal pathway. These spatial processes could be encoding spatial features such as colinearity of the local forms or they could be responding to the low spatial frequencies of the global forms (see Ivry & Robertson, 1998) or a variety of other spatial properties. Whatever the case, the global form was implicitly encoded but not explicitly accessible.

Implicit Spatial Cuing Effects

If implicit spatial representations were intact in Balints syndrome, the existence of these representations also addressed another puzzling finding, and that was that RM showed normal exogenous cuing effects in a Posner spatial cuing design (Egly et al., 1995). When two outlined boxes were placed on each side of fixation and one brightened as a cue but was not predictive of target location (exogenous cuing), RM was 84 ms faster to detect targets in valid locations than in invalid locations. Again, he could not report where the cues or targets were.

It might be argued that it was not the location of the brightening box that attracted his attention, but rather the bright box itself (i.e., the object). That is, he did not need to know where the brighter box was as long as the target appeared in a location within that box shortly after it disappeared. When the target appeared in an invalid box, he was slower to detect it than when it appeared in a valid box. However, RM could not explicitly report any better than chance whether two sequentially presented stimuli were in the same location or not, nor could he report whether a dot was inside or outside of a circle (Friedman-Hill et al., 1995;

Robertson et al., 1997). It therefore seems unlikely that object-based attention could account for his normal cuing effects.

A different but related argument could be that the cuing results were due to location priming, something that again would not rely on knowing where the cues or targets were located. A cue appears in the left or right visual field (the prime) and is followed by another flash (the target) either in the primed location (valid trials) or in an unprimed location (invalid trials). If implicit maps were present, then the location consistency between the cue and target should in itself speed response time while spatial inconsistency should slow it. Again, recall that RM was unable to report whether two flashes of light on a computer screen were in the same or different locations, yet he responded within a normal range when a cue and the target were in the same place. These effects are consistent with implicit spatial coding (at least in the sense that he implicitly encoded the correct spatial relationship between the cue and target). In normals, a peripheral flash of light automatically cues attention to its location (Posner, 1980). This appears to have been the case for RM and other Balints patients as well (Coslett & Saffran, 1991, Humphreys & Riddoch, 1993; Verfaellie, Rapcsak, & Heilman, 1990) even though the location of the target was not explicitly known.

We also tested RM with a centrally presented predictive arrow (endogenous cue) to examine his ability to volutarily select a location, but no cuing effects appeared. Although the dissociation between exogenous and endogenous cues was intriguing at first, we later found that he had difficulty perceiving the correct orientation of all types of objects, including arrows. The arrows were an ineffective cue, probably because their orientation was explicitly unknown, although evidence described later demonstrates that their orientations are implicitly encoded (Kim & Robertson, 2001).

Implicit Spatial Stroop Performance

Other studies designed to explore RM's implicit spatial abilities supply further support for intact implicit spatial maps as well as showing that these maps can be involved in processing far more complex spatial relationships than exist in cuing or priming studies discussed so far (Robertson et al., 1997; Kim & Robertson, 2001). In one experiment, a spatial Stroop task was used with the word UP or DOWN presented at the top or bottom of a rectangle (Figure 5.7), such that the meaning of the word was either consistent or inconsistent with its location. When normal perceivers are asked to read the word as quickly as possible, they produce slower reaction

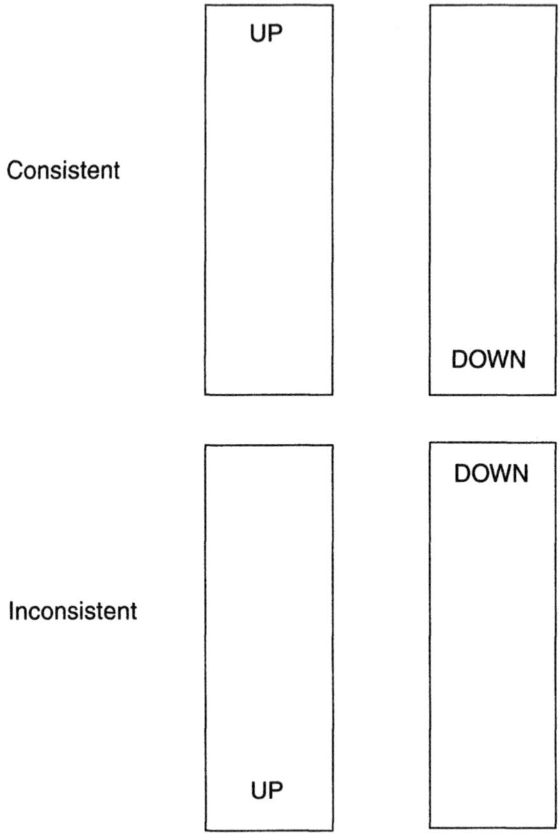

FIGURE 5.7. Example of spatial Stroop stimuli in which the word UP or DOWN is placed in a consistent or inconsistent location with its meaning. (Adapted from Robertson et al., 1997.)

times in inconsistent than consistent conditions (24 ms on average). Since RM could read individual familiar words (Baylis, Driver, Baylis, & Rafal, 1994), we measured his reaction time to read the word UP or DOWN across several blocks of trials in different sessions. Although his average spatial Stroop effects were larger than normal, Stroop interference was clearly present (142 ms).

In another block of trials using the same stimuli in free view, we asked RM to report whether the word was at the top or bottom of the rectangle, and his performance fell to chance levels. His inability to perform this task normally was not only reflected in his accuracy score (51%) but also in the discomfort he exhibited during testing. He would shake his head back and forth and protest that he did not know where the word was. He had to be prodded to guess the location.

We ran this experiment several times over a 2-year period during a time when RM was showing signs of some spatial recovery in his everyday life. At one point he was able to locate the words in the spatial Stroop task 83% of the time with a 4-second presentation (still not normal but substantially better). Although he did get somewhat faster at reading words over time, the spatial Stroop effect remained the same. While his explicit spatial abilities improved, implicit spatial abilities were unchanged.

There was also a period of time when his ability to locate the word returned to chance levels (49%). Three years after his second stroke and 2 years after we first tested his word location abilities in the spatial Stroop task, he suffered a spontaneous subdural hematoma (one of the very unfortunate side effects that sometimes occur with the anticoagulant medications he was taking to prevent further blood clots). This new event created a pocket of blood between his right frontal lobe and skull that increased cranial pressure. The blood was surgically evacuated and he was shortly transferred to a rehabilitation center. During this time, his spatial problems returned to the level that we saw upon initial examination some years earlier along with all Balints symptoms. A few months later he had recovered visual spatial function to the levels we observed just prior to the subdural (as the pressure on his brain subsided), and there was no radiological evidence of residual mass or new or extended lesions. During a short period after the hematoma RM unfortunately lost all the gains he had made. Even with this setback, he requested the nursing staff to contact us to continue the research. This created a situation that is as close as neuropsychological investigations ever come to an ABA design.

During this period we reran several tests RM had performed earlier. These included the spatial Stroop tasks and others that will be discussed in the next chapter. But for the purposes of examining implicit space, the variations in his explicit spatial abilities and his spatial Stroop effects are the most informative. Again, although he was at chance levels in locating the words UP and DOWN in a rectangle, he was faster at reading the words when they were consistent with their location than when they were inconsistent. Furthermore, the magnitude of these spatial effects was not significantly different from those we observed on any previous occasion. These findings provided further evidence that explicit spatial abilities could not account for the implicit spatial effects. They also raise questions about explanations based on thresholds or response biases of the previous findings. Implicit effects remained relatively stable over wide fluctuations in explicit spatial abilities. Notice that the word itself was explicitly perceived by RM. The semantic information was explicitly processed, allowing the semantics of the word to have a top-down influence on a spatial system represented below the level of awareness.

Implicit Localization of Complex Spatial Relations

It is not news that stimuli presented in areas of space of which patients are unaware can affect performance. The now rather extensive evidence in the hemineglect literature has shown that even semantic information can be encoded without awareness that a word was present at all (Ladavas et al., 1993; McGlinchey-Berroth et al., 1993). Yet the common denominator among all visual stimuli, whether written words or objects, is that they themselves are spatial structures that can be defined by object-centered reference frames. Some spatial processing is necessary for implicit or explicit object-based effects to emerge. There must be some spatial map(s) that carry information without the explicit spatial processing of the parietal lobes.

In a recent study we demonstrated that RM could implicitly represent the location of more than one object below the level of awareness, showing that even with bilateral damage spatial representations that do not require parietal input include complex spatial relationships between separate items (Kim & Robertson, 2001). We used a dual task procedure where subjects first responded as rapidly as possible to the presence or absence of an arrow that appeared at fixation. They then reported whether a unique feature had been present or not in a multi-item array that had appeared briefly just prior to the onset of the central arrow (Figure 5.8). On each trial subjects first fixated a star-like pattern that appeared in the center of a computer monitor. This pattern was followed by a 60 ms exposure of a feature search display with four circles in the periphery equidistant from fixation. Either 60 or 300 ms after the search display disappeared, the central fixation changed to an arrow that either pointed in one of four directions (toward one of the circles that had briefly appeared in the search display) or changed to a symmetrical pattern. The subjects' task was simply to push a button as soon as they detected an arrow regardless of its orientation and to withhold responding if it changed to one of the symmetrical patterns instead. After their response to the arrow or after a set interval, subjects were asked whether a target (e.g., a red target among green distractors) had appeared in the search display on that trial.

Notice that spatial information was not necessary to perform either task. Neither the location of the target in the search display nor the orientation of the arrow was relevant (although normal perceivers almost certainly perceived both). The first task was simply to push a button when an arrow appeared no matter which way it pointed, and the second task was to say whether or not a feature target had appeared in the search display. The pattern of reaction times to detect the arrow was the main interest, and for normal perceivers and RM this pattern was similar.

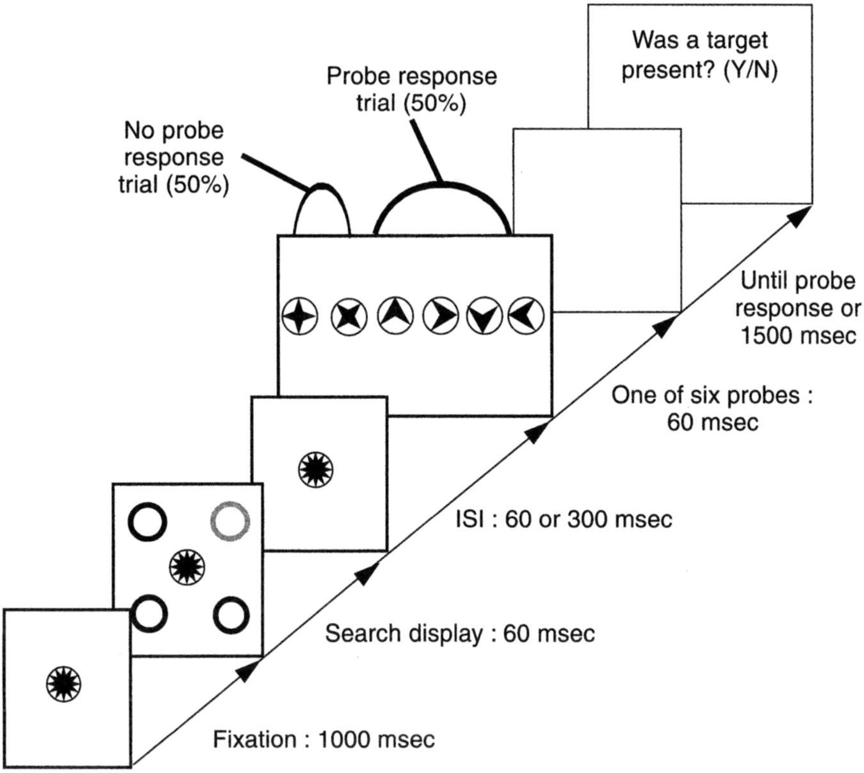

FIGURE 5.8. Schematic of a trial sequence used to test for implicit spatial information with RM. The search displays contained one red and three green or three red and one greed circle when a target was present. When it was absent the circles were all the same color. After the search display disappeared, the fixation point reappeared and either 60 or 300 ms later changed into one of the four arrows or two symmetrical probe shapes. Instructions were to respond to the probe as rapidly as possible if it was an arrow and to withhold response if it was not. Later in the trial, participants were asked whether a target appeared in the search display or not. Note that no location information was required to perform either task.

When the arrow pointed to where a feature target had just appeared in the search display, responses to detect an arrow were fast, as would be expected if attention had been drawn to the location of the target (Kim & Cave, 1995). But what was most interesting was that the speed of detecting the arrow increased linearly over orientation from the feature target location. Reaction times to detect the arrow were slower when the arrow pointed toward a location horizontal or vertical from where the feature target had appeared (+/–90°) and slower still when it pointed toward the location opposite to where the feature target had appeared (180°). Reaction time to detect the arrow increased with the orientation between the arrow and the location of targets and distractors in the display (Figure 5.9a). This reaction time pattern in normal perceivers was more pronounced at shorter interstimulus intervals (ISIs) than at longer ISIs. However, there was no significant interaction between delay and the orientation distance.

Although the question of why this pattern might occur is interesting in its own right and needs further study in order to specify the underlying mechanism (e.g., captured attention, spatial coding, etc.), it is clear that the spatial relations between where the feature target appeared a few milliseconds before an arrow and the arrow's orientation was encoded even though no location or orientation information was needed to perform the task. Notice that processing the spatial relationship between the arrow's orientation and the feature target as well as the spatial relationship between the arrow and the degree of misalignment with the distractors in the search display would be required in order to produce the linear pattern observed.

But the most important question for the topic at hand was whether a person who had virtually no explicit spatial information would show a similar pattern when detecting the central arrow, and indeed RM did (Figure 5.9b). As with normal perceivers, he was told that he was to look at the center, where the star-like fixation pattern appeared. As soon as it changed to an arrow he was to respond as fast as he could by pressing a response button and refrain from response if it changed into any other pattern. After responding to the central pattern, he was then asked whether or not the feature target had appeared on that trial. RM was very good at detecting the arrow, as were normal perceivers (93% vs. 98%, respectively). Since the arrow appeared by itself on the screen after the search display disappeared and was presented where RM was fixating, arrow detection was quite easy for him.

One way in which RM's data differed from normal subjects was that he missed a significant number of feature targets (72%) at these brief displays (normal subjects missed less than 2%). When his attention was centered on the fixation pattern, peripheral features were not well detected. (RM's feature detection performance was likely so poor because he was asked to report the feature target only when he had high confidence that he saw it.) The data shown in Figure 5.10 thus reflects cases when he was sure he saw it. Nevertheless, the locations of the target were explicitly unknown. He showed evidence of implicit spatial encoding of the spatial relationships between a central arrow and peripheral target as well as between the arrow orientation and degree of angular disparity from the target. Implicitly, targets and distractors and their spatial relationships to the arrow's orientation were encoded, despite the fact that the orientation of the arrow and the location of features were irrelevant for the task.

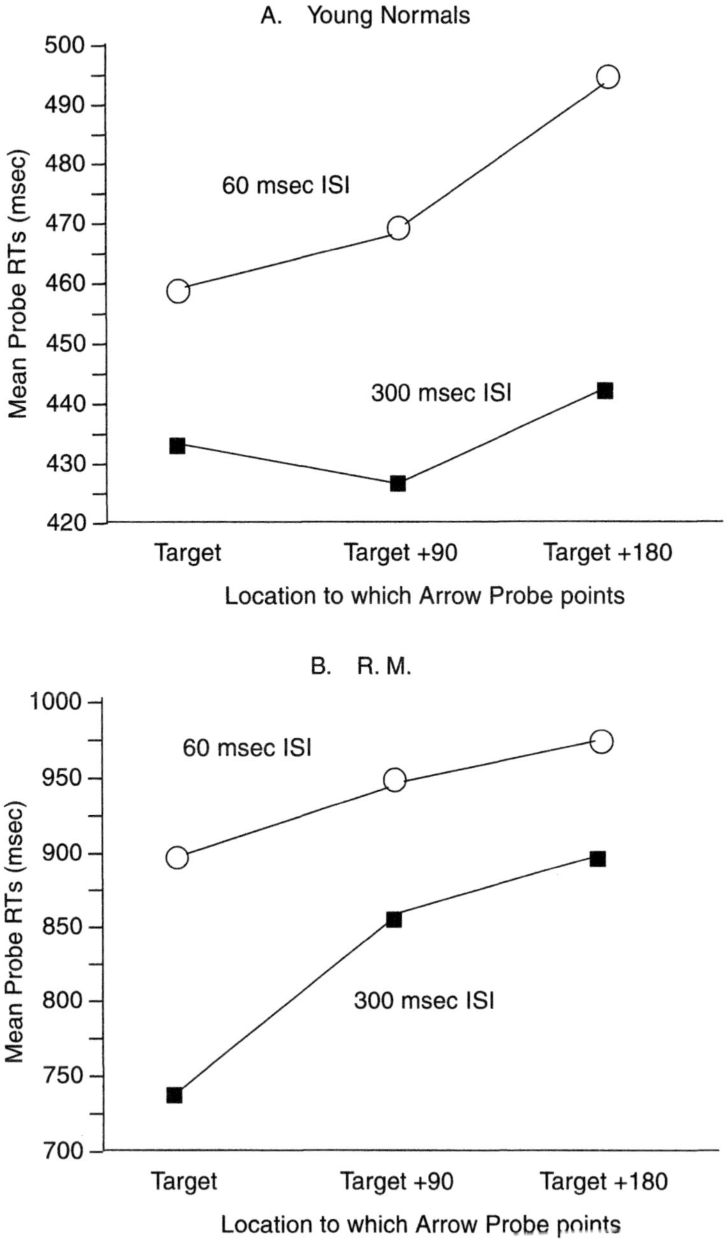

FIGURE 5.9. Mean reaction times for young normals (a) to detect the presence of an arrow as a function of whether the arrow probe pointed to the location where the search target had appeared, to one of the distractor locations 90° away from the search target or to the distractor location 180° away from the search target for the two interstimulus intervals (ISIs). Mean reaction time collapsed over eight sessions for RM plotted in the same way (b). (Adapted from Kim & Robertson, 2001.)

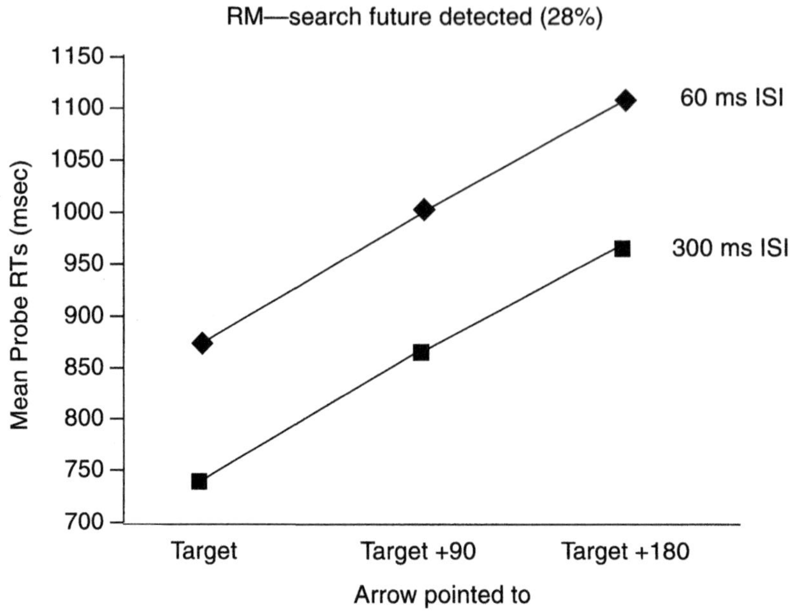

FIGURE 5.10. Mean reaction time to detect the arrow for RM plotted in the same way as in Figure 5.9 for hit trials only (when he reported seeing the search target). Note that he was 93% correct in detecting the arrow itself, and only correct arrow detection responses were included to calculate mean response time.

To make sure that RM could not explicitly determine these spatial relationships at this time, we performed another study in which we presented stimulus displays for several seconds that included both the four-item search display and the central arrow at the same time. The arrow pointed either to the target or to one of the distractors (Figure 5.11). We asked RM to say "yes" if the arrow pointed to the target and "no" if it did not. His overall bias was to say yes, but his yes responses were nearly the same when the arrow pointed to the location of the target as when it pointed to the location of one of the distractors (62% and 61%, respectively). He clearly did not explicitly know the spatial relationship between the search target and the arrow, yet his reaction time performance in the first experiment indicated that he did process this information.

From the several studies I've discussed in this and previous chapters (using different stimuli in different types of tasks) there is a great deal of evidence that rather complex spatial information is represented even in the face of severe disruption of dorsal spatial function. The source of the implicit spatial information is not yet known, but the evidence for such a high level of spatial description may help answer how it is possible that a single object can be perceived without explicit spatial knowledge. The

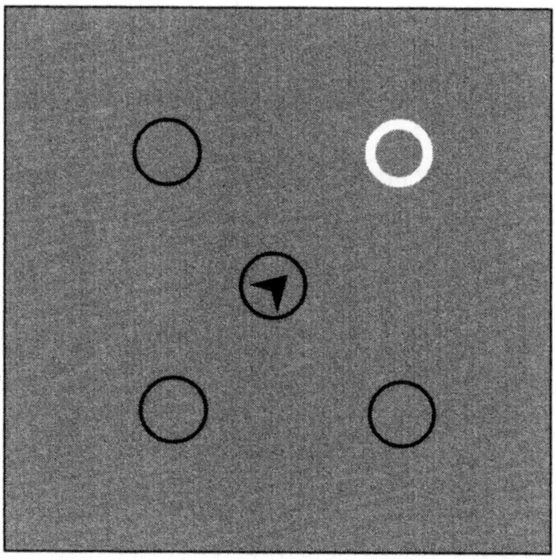

FIGURE 5.11. Example of display shown in free view with the arrow pointing to a white circle (the circles in the display were actually red and green; the black here represents red and the white represents green. The arrow and its surrounding circle were black as shown). RM's task was to say yes if the arrow pointed to a red circle and no if it pointed to a green circle. He said yes as often when it pointed to a red as when it pointed to green circle.

major question revived by the Kim and Robertson (2001) results demonstrating implicit spatial information is why more than a single object is not perceived if the spatial array is so well represented in the visual system. Preattentive visual processing that binds lines, angles, and volumes into a spatial configuration we call an object appears to be working well enough for RM to identify an object. In addition, the spatial relationships between "objects" are also encoded but are unavailable for spatial awareness. Awareness is also blocked for more global levels of perceptual organization that require grouping across spatial gaps. It is the explicit access to spatial relationships between individual units that seems to be the province of the posterior dorsal processing stream.

☐ Functional Aspects of Dorsal and Ventral Processing Streams Reconsidered

It would be hard to overestimate the influence of Ungerleider and Mishkin's (1982) proposal that a dorsal stream determines "where" things are, and a ventral stream determines "what" things are. A great deal of evidence has confirmed parietal lobe functions in processing space and the temporal lobe functions in processing objects. Yet the findings discussed in the last section demonstrate the existence of a great deal of spatial processing below the level of awareness even when both parietal lobes are damaged.

The intact parts of RM's brain were sufficient to encode complex spatial relationships but were not sufficient to bring that information to awareness. If the dorsal system is in charge of processing where things are, how does this occur?

A proposal by Humphreys and Riddoch (1994) is intriguing as it relates to this question. They suggested that the dorsal and ventral streams could be distinguished by the different types of spatial information that each computes. They hypothesized that spatial relationships between objects were processed by the dorsal stream, while spatial relationships within objects were processed by the ventral stream, and noted that this model could explain why Balints patients can see objects at all (albeit only one at a time). The intact ventral system would be sufficient to represent the spatial structure that supports perceptual awareness of an object, while the dorsal system would be necessary to represent spatial relationships between objects and thus support awareness of more than one object in a visual scene. However, their arguments were meant to account for explicit abilities of patients with dorsal stream damage. The evidence derived from RM demonstrates that spatial information beyond a single object is encoded even when it cannot be reported, so something more is necessary.

In the arrowprobe task described in the last section, the central arrow was a different "object" from the colored circles that were placed around it, and a red target was a different "object" than a green distractor (or vice versa). Nevertheless, the spatial relationships between these objects were encoded correctly by cortical systems outside the areas of parietal damage, potentially with a major contribution from the ventral processing stream. One question, then, is to what extent spatial information within an object can be explicitly selected for awareness. That is, once an object is perceived, can the locations of the parts be explicitly attended? Again, the strongest tests have been derived from patients with Balints syndrome and bilateral parietal damage which would deprive the ventral stream from input from posterior dorsal stream processing.

Ventral Space and RM

To address this issue, Anne Treisman and I again called upon RM. We presented rectangular stimuli horizontally or vertically that were either connected to form a closed figure or unconnected by the presence of a gap (Figure 5.12). On some trials the stimulus had a curved end rather than a straight line. RM's task was first to detect the presence or absence of the curve by saying yes or no and, if he said yes, to report whether the curve was on the left, right, up, or down. The stimuli were always centered in the middle of the screen, so the spatial locations of the curve also coincided

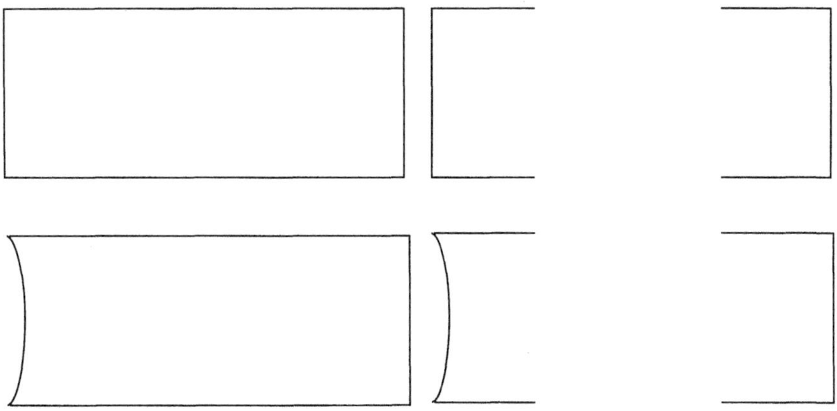

FIGURE 5.12. Example of stimuli used to test RM's ability to detect a curve in a closed (left) and open (right) figure. He first reported if a curve was present or not and, when he said it was present, whether it was on the right or left of the display.

with the spatial locations in both screen-based and viewer-based coordinates. Any of these could be used to perform the location task, and in this way we hoped to maximize the availability of useful spatial information.

Consistent with previous results in Balints patients (see Luria, 1959; Humphreys & Riddoch, 1993, Humphreys, Romani, Olson, Riddoch, & Duncan, 1994), we found that closure affected performance. RM was 90% accurate at detecting the curve when the stimulus was connected and 75% when it was not. This level of accuracy occurred whether the rectangles were presented for a few milliseconds (197 ms) or for several seconds (up to 5 seconds). The pattern of performance was consistent with the Balints symptom of simultanagnosia. In contrast, RM's ability to locate the curve was poor, and there were no differences between his localizing abilities when the stimulus was connected and when it was not (both 64%).

These findings demonstrate that closure is not more likely to bind objects to a location but does increase perceiving stimuli as bound wholes. They also support previous findings with Balints patients that perceiving a feature and knowing its location are dissociable (see De Renzi, 1982; McCarthy & Warrington, 1990).

Note that in our study location judgments were only made after RM responded that a curve was present. There were only a few trials when he responded that it was present when it was not (false alarms), so the vast majority of his curve detection errors were misses. Given his simultanagnosia we can assume that for displays with gaps, he saw either the portion that contained the curve or the portion that did not. For connected displays he saw the one object in the display and was therefore more accurate in judging the presence or absence of the curve. In either case, he had great difficulty in locating the position of the curve on the screen.

These results suggest that RM's location errors could be attributed to a deficit in evaluating the frame of the object relative to a more global frame or to the intact frame of his own body. Without the ability to bind the object frame to another frame, the location of the curve could not be known even if the relative locations within the object frame itself remained constant but transformed (e.g., through rotation). So we devised another way to test RM's ability to explicitly perceive the relative location of parts within an object.

To accomplish this end we used words and letters in which the spatial relationship between two parts was the only factor that defined their identity. For instance, the words NO and ON are only discriminable by explicitly seeing the relative locations of the O and N. We showed one of these words per trial in the middle of a blank screen and in some blocks of trials asked RM to report whether the letter N was on the left or right. He found this task very difficult, took a long time to respond, and was at chance performance. However, when we asked him to read the word NO or ON in other blocks of trials he was 69% accurate (clearly not good, but significantly better than when his task was to locate the N). Although he could not explicitly access the location, there was evidence of some implicit encoding of spatial information that influenced the identification of the word.

But if he explicitly saw the word NO, why couldn't he correctly report that the N was on the left? We speculated that his perception of the relative location of the letters during word reading was helped by the familiarity of the words NO and ON. To test this hypothesis we replaced the words ON and NO with the less frequent letter strings OZ and ZO. We hypothesized that this would decrease the influence of top-down mechanisms because the semantics of these letter strings would not be accessed automatically. As predicted, RM was no better than chance at either reading the word or localizing one of its letters when the letter strings were less familiar. Although the amount of data we were able to collect in this study was limited for practical reasons, these findings suggest that the more familiar the word was the more top-down information influenced his ability to perceive the word as its proper whole.

Ventral Space in Another Balints Patient

In what at first seemed in direct contrast to our findings with RM, Cooper and Humphreys (2000) reported that another patient with Balints syn-

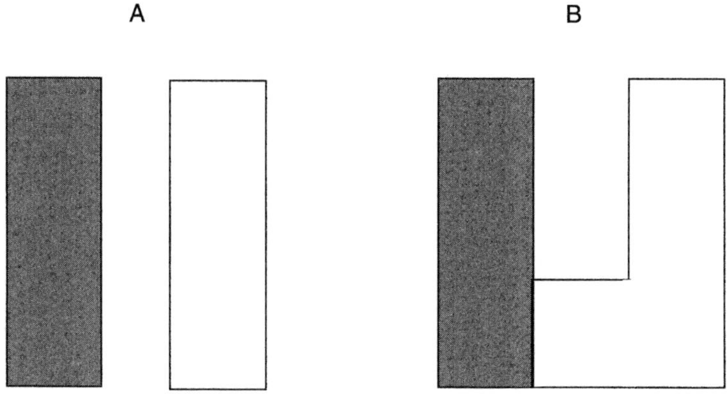

FIGURE 5.13. Example of stimuli used to test a Balints patient (GK). He was asked to report whether the gray (actually colored) bar was on the right or left when the bars were separated (a) and when they were connected (b) (Adapated from Cooper & Humphreys, 2000.)

drome (GK) was better at locating parts within an object than locating one object relative to another.[4] They presented black and red vertical bars to the right and left of each other (Figure 5.13a) for 3 seconds and asked GK to report where the red bar was located relative to the black bar. He was correct only 43% of the time. However, when the two bars were connected to form a U shape (Figure 5.13b), performance significantly improved (88%). Cooper and Humphreys concluded that locating parts within an object utilizes a different spatial system than locating objects relative to one another, supporting Humphreys and Riddoch's (1994) claims for different dorsal/ventral spatial functions.

Again, these results are inconsistent with our findings that RM was as poor at locating a curve within a connected figure as in two figures separated by a gap. Although detection of the curve was affected by closure for RM, locating the curve was not. In another study, Cooper and Humphreys (2000) asked GK to report whether the bars in each of the patterns in Figure 5.14 were the same or different heights, and he was no better than chance (54%) for Figure 5.14a, while he was 86% correct for patterns shown in Figure 5.14b, where colinearity of the lower horizontal line was magnified. In fact, the colinearity in Figure 5.14b produced similar results to when the figure was a whole closed shape as in Figure 5.14c (84%). When the colinearity of the base was disrupted as in Figure 5.14d, accuracy decreased. However, notice that in both b and c, the patterns with different heights begin looking like a J and those with the same heights begin looking like a U.

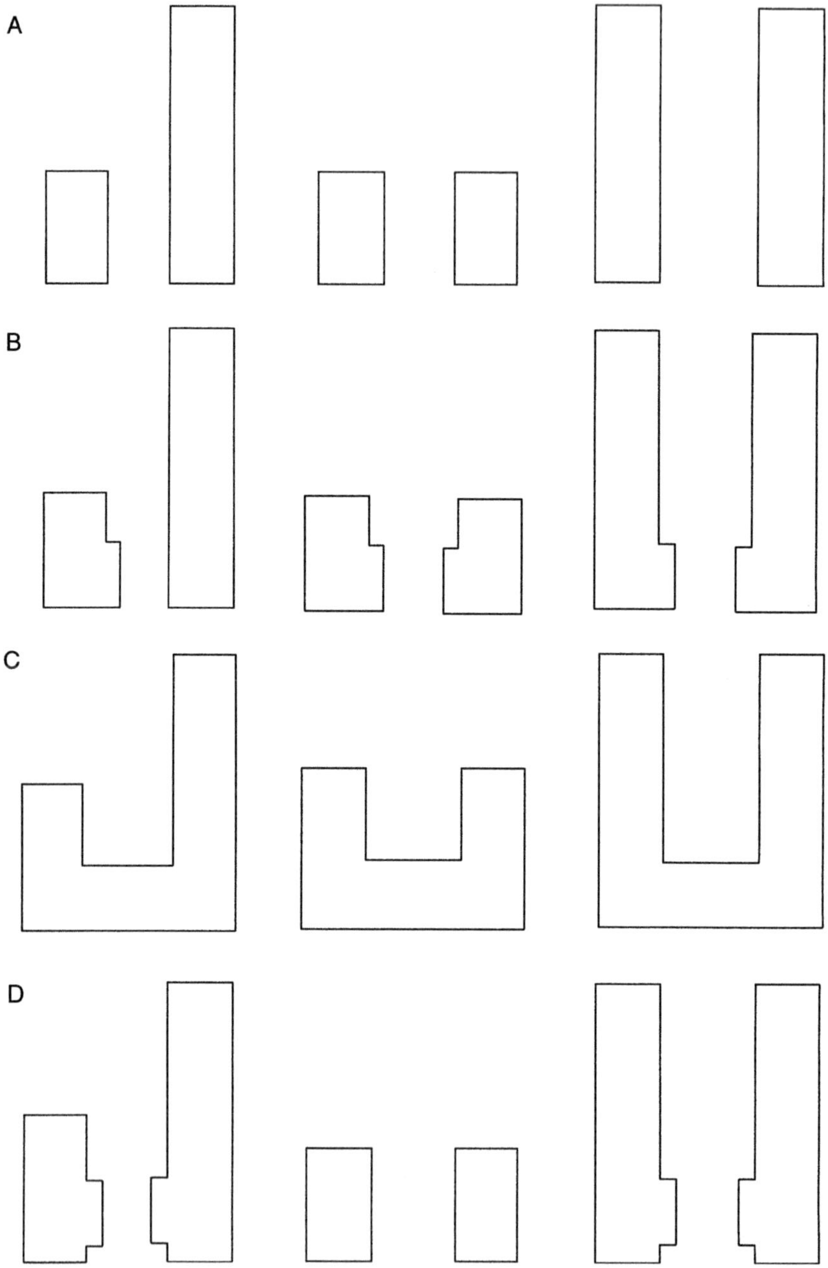

FIGURE 5.14. Examples of stimuli used to test the role of closure and collinearity with patient GK. He was asked to report whether the two elements in each stimulus were the same or different heights. He was better at this judgment in (b) and (c) where the patterns looked more like a J or U, than in (a) and (d).

Cooper and Humphreys (2000) did in fact make this observation and concluded that this made the figures in 5.14b and 5.14c more likely to be processed by the ventral object-based system. They further argued that spatial information within objects was *explicitly* available through ventral processing, and again concluded that the dorsal stream was used to direct spatial attention between objects, while the ventral stream was used to direct it within objects.

It would make theoretical development much easier if this simple division were true, but it turns out that the story is far more complicated, as is so often the case in scientific endeavors. A more recent paper reporting data from the same patient found effects that were not consistent with this conclusion using different stimuli and, in fact, were in the opposite direction (Shalev & Humphreys, 2002). GK was better at between-object than within-object localization. For instance, when GK was asked to locate the gap in Figure 5.15 by saying whether it was toward the top or bottom of the vertical line (i.e., the object), he was only 52% correct, or basically at chance. However, when he was asked to report whether the short line was above or below the longer line (a between-object task) he was 79% correct.

Inconsistencies Resolved

Before you throw up your hands and say, "I've had enough of these single case studies in neuropsychology and their inconsistent results," let me

FIGURE 5.15. Example of a simple line stimulus used to test GK's ability to use objects to make location judgments. In one condition he was asked to judge whether the gap was at the top or bottom of the line (within-object condition) and in another he was asked to judge whether the small line was above or below the long line (between-object condition). Unexpectedly, he was better at between-object judgments than within-object judgments. (Adapted from Cooper & Humphreys, 2001.)

assure you that the picture becomes quite clear with closer inspection. Both Shalev and Humphreys (2003) and Robertson and Treisman (in preparation; see also Robertson et al., 1997) concluded that ventral processing for locations within objects benefited from top-down information, and the more familiar an item was, the more top-down processing there would be.

Like other Balints patients, both GK and RM could identify single objects. With RM's intact ventral pathway, one might expect that the spatial relationships that define the shape of a single object would become explicit. However, RM had problems perceiving certain spatial properties of even a single object. He sometimes reported seeing a normal face when a jumbled face was presented and he reversed the order of letters within a word when the letters could produce more than one acceptable word (e.g., TAP and PAT). His reliance on top-down information to recognize these stimuli suggests that explicit spatial relationships within objects may not be intact either.

Shalev and Humphreys (2002) resolved their seeming inconsistencies with GK in a clever way that turned out to be more like RM than at first appearance. They presented stimuli like that shown in Figure 5.16 and first asked GK to report whether the pair of smaller circles were at the top or bottom of the oval in a series of trials. There was little evidence for within-object localization that was any better than chance (55%). They

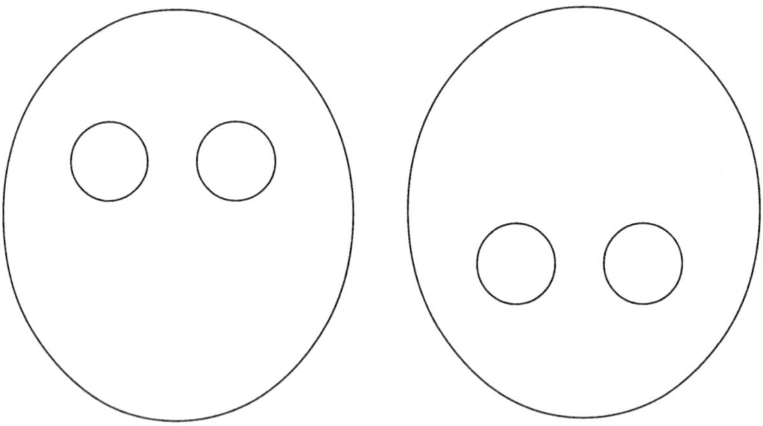

FIGURE 5.16. Examples of stimuli used to test the role of instructions to influence top-down processing by GK. In one condition he was asked to determine whether the pair of smaller circles were at the top or bottom of the larger oval. In another condition he was told that the pair of circles were eyes in a face and he was asked to report whether the eyes were toward the top or bottom of the oval. When the stimuli were primed as faces, he was much better at the task than before the face instructions were given. (Adapted from Shalev & Humphreys, 2002.)

then presented the same stimuli but asked GK to report whether the eyes were at the top or bottom of the face, and his performance dramatically improved (91%). When he thought of the stimuli as faces, he was able to determine whether they were upright or upside-down (see Footnote 1). Shalev and Humphreys went on to determine whether this improvement was from top-down influences alone or due to some interaction between top-down information and bottom-up perceptual cues by showing the stimuli represented in Figure 5.17. The instructions were to report the location of the "eyes" in the face, but now the eyes did not look much like

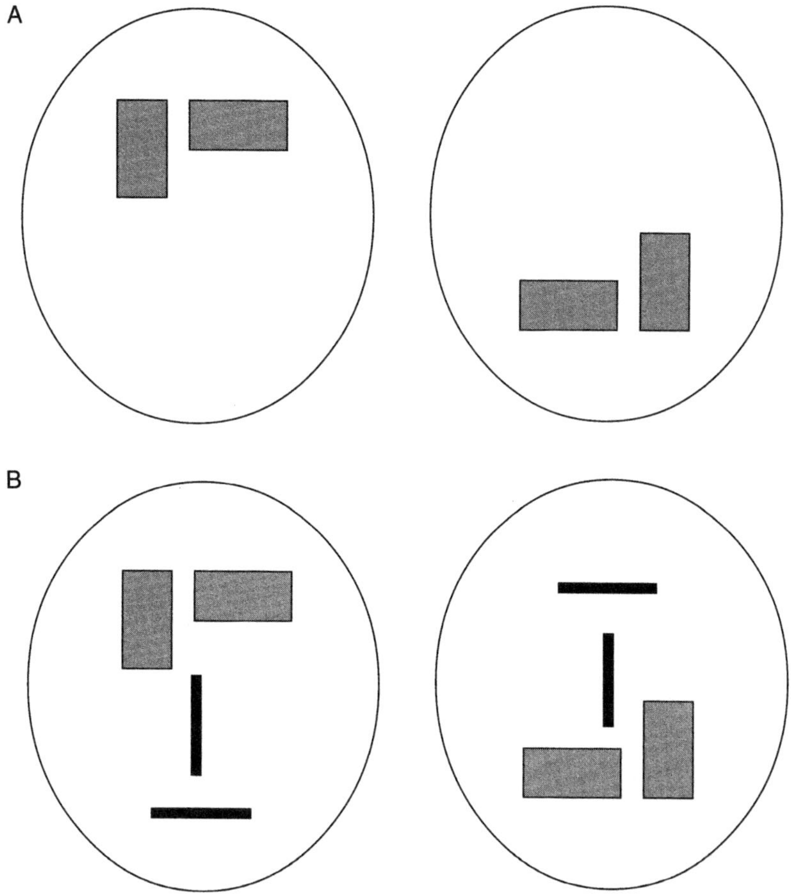

FIGURE 5.17. Stimuli used to test how perceptual features interact with top-down processing with patient GK. He was told that the two rectangles were eyes in both cases a and b, but now he was better at judging their location when they were accompanied by lines denoting a mouth and nose than when they were not. (Adapted from Shalev & Humphreys, 2002.)

eyes and only their locations defined them as such (Figure 5.17a). They then added additional features to make the eyes look more like eyes (Figure 5.17b). In the eye-unlike condition, GK was again very poor at localizing the "eyes" (55% correct), but when the perceptual cues were added to make the "eyes" look integrated into a facial configuration, location performance improved to near perfect (98%). GK's ability to determine whether the faces were upright or upside-down was good as long as the bottom-up information was sufficient to stimulate a match to top-down information. When the "eyes" were positioned so that the face matched the internal upright representation of a face (the canonical orientation), the location of the eyes would be toward the top. When the eyes were positioned toward the bottom (normally signaling an upside-down face), the location of the eyes would be toward the bottom. In other words, the location task could be done successfully by using impressions of whether the face was upright or upside-down, but this was a combination of top-down and bottom-up processing.

This influence of top-down information on perceptual processing was also addressed by Elizabeth Warrington and her colleagues some years ago (see Warrington & Taylor, 1978). On the basis of studies of patients with unilateral damage to the right or left posterior cortex, she suggested that the right hemisphere was involved in perceptual categorization, and the left hemisphere in semantic categorization of objects. Patients with right hemisphere damage had difficulty matching photographs that differed in perspective, leading her to suggest that viewpoint invariance of the object (i.e., an object-based reference frame) was associated with the right hemisphere. When damaged, the frame of reference no longer afforded location information needed to know what the object was *unless* there was top-down information. In the model Warrington and Taylor proposed this top-down information originated from verbal memory. Although these findings were interpreted with regard to hemispheric differences, they also demonstrate a dissociation between top-down semantic influences and bottom-up spatial processing.

□ Many "Where" Systems

One conclusion that can be made on the basis of the patient work discussed in the previous sections is that the "where" functions of the inferior parietal lobes are involved in explicit spatial knowledge, while they have only limited, if any, role in implicit spatial encoding. But where does this leave us in terms of knowing where the sources of implicit spatial effects arise? Most of the evidence bearing on this issue has come from animal work, and there are several candidates. The primate brain has many "where" systems, some of which would remain intact even with large bilateral parietal lesions

Evidence from single unit recordings in monkeys supports the patient literature and also suggests the existence of multiple spatial reference frames. Some spatial maps can be represented below the level of awareness, since neurons in many areas are as active when animals are awake as when anesthetized. It is not my intention here to review this vast literature, since the focus in the animal literature has not been on the relationship of these maps to spatial awareness per se or on top-down influences on spatial representations. Nevertheless, there are several relevant studies that bear on the proposals I've been discussing, namely the representation of multiple spatial reference frames, and I will highlight some intriguing findings that are at least consistent with the patient data I've discussed throughout this book.

Before beginning, it is important to point out that the evidence for multiple spatial maps has been available for some time. There are spaces that support action and others that support perception (Colby & Goldberg, 1999, although for a singular view see Rizzolatti, Berti, & Gallese, 2000). Different maps seem to govern personal versus peri-personal versus extra-personal responses (see Bisiach, Perani, Vallar, & Berti, 1986, Rizzolatti, Gentilucci, & Matelli, 1985). There are also many reported dissociations between viewer versus retinotopic versus extrapersonal spatial representations (see chapter 3). However, the idea of spatial maps that are hierarchically organized within each of these systems has not been straightforward, and the distinction between implicit and explicit spatial maps has not been a concern. Understanding implicit maps seems especially important when trying to articulate what explicit spatial representations might remain when damage to the brain causes spatial deficits that disrupt everyday life. In addition to their value in understanding brain function, finding a way to access remaining spatial maps could prove quite valuable for cognitive and visual rehabilitation programs.

By now it should be obvious that parietal damage in humans is likely to result in spatial deficits, and these deficits can take many different forms. Some patients lose their sense of body space: a patient with left hemineglect might push her own left arm away as if it is an intruder (Brain, 1941). Other patients may report sensation on their left side as coming from their right (allesthesia), while others may ignore stimulation on the left entirely (Heilman, Watson, & Valenstein, 1993). Some individuals have exhibited neglect in near but not far space and vice versa (Cowey, Small, & Ellis, 1994; Halligan & Marshall, 1991). A subset of patients with neglect show evidence of motor neglect but not perceptual neglect, while others show the opposite (Mesulam, 1981). Although motor neglect has been most often associated with lesions adjacent to motor cortex, a recent study by Ro, Rorden, Driver, and Rafal (2001) demonstrated that parietal lobe lesions could disrupt saccadic eye movements while not affecting visual encoding. These examples from patient studies are consistent with the idea that the parietal lobe itself contains several different spatial maps.

A recent influential proposal is that the parietal lobes coordinate various other systems in distributed areas that are spatially specialized (Gross & Graziano, 1995). In other words, the parietal lobe acts as a control center for spatial selection. According to this view, the various constellations of spatial deficits after parietal damage are due to disconnections between different regions of the parietal lobe and other spatially sensitive areas (Figure 5.18). Some examples of spatially sensitive areas within the brain with strong connections to the parietal lobe have been thoroughly reviewed before (Colby & Goldberg, 1999; Gross & Graziano, 1995) but it is useful to briefly describe here the major ones that may contribute to implicit spatial effects observed in Balints patients.

Somatosensory-Visual Bimodal Cells

Somatosensory-visual cells respond both to a body part and to visual stimuli, making them prime candidates for eye–hand coordination, a task that requires fine spatial precision. One of the first reports of these cells focused on the ventral premotor cortex (PMv) of the frontal lobe (Rizzolatti, Scandolara, Matelli, & Gentilucci, 1981). The PMv contains a topographical map of the body, and these bimodal cells are not linked to locations on the retina or to locations in the world. Rather, they are tied to the receptive field of a body part (Fogassi et al., 1992). For instance, when the hand is in view, the visual response follows the arm rather than staying fixed at a given location relative to the retina. The reference map in this system is body space, not retinotopic or environmental space.

Similar somatosensory-visual bimodal cells also have been found in the putamen, which contains a rough topography of the body (Graziano & Gross, 1993). Again, a cell's receptive field follows the body part rather than the eye, similar to the cells in PMv (Gross & Graziano, 1995). Some cells respond to body parts that seldom come into view, such as cells centered on areas of the face. However, these cells do respond to visual stimuli presented a small distance from the face where they would be visible. Here the receptive field for visual stimulation follows the movement of the face, which of course follows the movement of the head.

Since frontal and subcortical areas can represent such spaces, then why does parietal damage affect the ability to determine the location of an item and reach for it correctly? Gross and Graziano (1995) suggested that

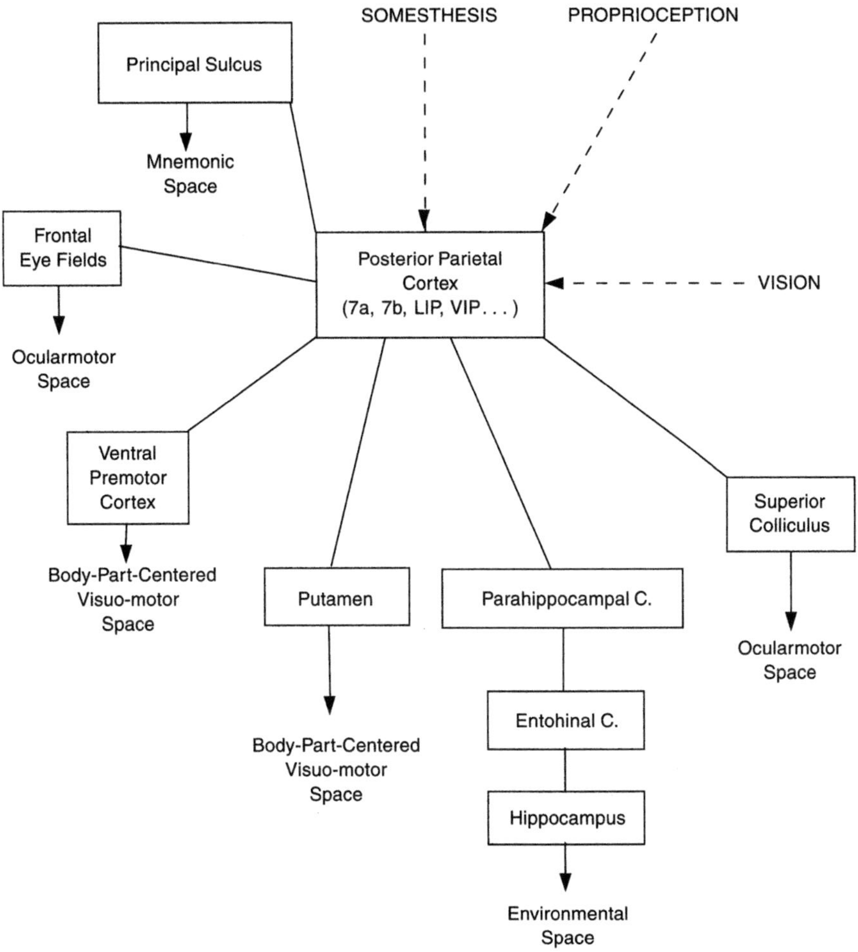

FIGURE 5.18. A schematic of the different areas of the brain that contain spatial maps and connect to posterior parietal cortex as mapped out by Gross and Graziano, 1995. (Reprinted with permission.)

parietal lobes represent little topography themselves, for either vision or touch, but are strongly connected to both PMv and putamen, which do have adequate topography. In their model, parietal damage would disconnect the parietal lobe from the putamen and PMv, producing some of the deficits like those in Balints patients. Parietal lobe damage to areas connected to the putamen and PMv should affect the ability to reach for objects as well as eye–hand coordination, and these are among the spatial problems seen in patients with parietal lobe lesions even when unilateral (De Renzi, 1982). An area of the parietal lobe that may be instrumental in this activity is in the medial intraparietal region (MIP). MIP contains neurons that are sensitive to space within reaching distance of the arm and respond to the location of a stimulus and its direction of motion.

It is also of interest to note that the bimodal cells of PMv, discussed at the beginning of this section, respond in anesthetized monkeys in the same way as in awake and behaving monkeys, meaning that spatial awareness is not necessary for space to be encoded within these maps. If these findings are applied to Balints syndrome, it suggests that the space encoded within these regions may support the implicit effects in these patients, but that the maps in this area are not sufficient to arise to awareness without parietal interaction. Certain parietal-frontal-subcortical connections appear to be necessary for individuals to become aware of the spatial relationships between a body part and a visual stimulus.

Ocularmotor Responses

Other areas of the brain are principally involved in ocularmotor programming, which requires a spatial map of the eyes in their sockets and the location and direction they must move. A spatial map that directs eye movements may well be represented in polar coordinates to facilitate the movement itself. However, horizontal and vertical axes are of special importance, as shown by the fact that saccades along these axes can be independently affected, as can be observed after damage to midbrain structures and in the early stages of some progressive dementias that begin within the midbrain (Rafal et al., 1988). The major areas that contain topographic mapping for eye movement control are the SC and FEF. In monkeys, areas within the lateral inferior parietal lobe (LIP) contain cells that respond to eye movements as well, and there are strong connections between all of these areas (Andersen, Essik, & Seigel, 1985; Cavada & Goldman-Rakic, 1989; Lynch, Graybiel, & Lobeck,1985).

The receptive fields of cells in all three of these areas are retinotopically mapped. That is, when the eye moves, the receptive field of the cell moves with it. This does not mean they are simple slaves to the retina by any means. Some LIP neurons fire in anticipation of a stimulus coming into their receptive fields when a saccade is planned (Goldberg, Colby, & Duhamel, 1990). Some LIP neurons also have memory for the spatial location of a previously shown stimulus. They will fire when brought into alignment with a location where a target has disappeared (Duhamel, Colby, & Goldberg, 1992). In such a case the receptive field of a neuron has not been stimulated from an external source because the stimulus is outside the receptive field when presented and is gone before a saccade is made. In addition, LIP ocularmotor neurons are modulated by the combination of eye, head, and body orientation (Andersen et al., 2000). In this manner, LIP neurons are sensitive to extra-retinal spaces, although eye-centered coordinates define the primary reference maps for this system.

It is important to note that LIP neurons have different response properties than neurons in other ocularmotor areas (Colby & Goldberg, 1999), although they may work in coordination with them. It is somewhat controversial whether LIP neurons are for directing attention or for intention to make a motor response (see Andersen, et al., 2000; Colby, 1996), but they do respond when a visual stimulus appears and some do so whether the monkey is trained to make an eye movement to a stimulus or not.

It also should be noted that another area within the anterior portion of the intraparietal sulcus (AIP) responds to object shape in a way that is useful to form the spatial configuration of the hand when grasping for an item. This area is also strongly connected to the premotor cortex and supports models of parietal function as basically a premotor area for action (Milner & Goodale, 1995; Rizzolatti et al., 1994). Although some areas of the parietal lobes are clearly involved in action, one must be cautious to conclude that the spatial maps associated with the parietal cortex are exclusively involved in representing space for this purpose (see chapter 6).

Primary Visual Cortex

Entry-level visual cortex contains a spatially isomorphic representation of the 2-D picture that is projected to the eye. The spatial map is detailed and precisely represents luminance contrasts in the stimulus as it is reflected on the retina. When visual cortex is ablated, blindness ensues. However, the phenomenon of "blindsight" has demonstrated that detection of light presented in the blind field can occur above chance, although individuals are unaware that a stimulus has been presented. Blindsight demonstrates that detection of a feature such as a bright light or color can disappear from awareness (Weiskrantz, 1986), but what happens to the location of that stimulus? Is awareness necessary in order to locate a feature that has not been seen? The answer seems to be no and is consistent with the implicit spatial effects that I discussed earlier in this chapter. Patients with blindsight can be up to 95% correct when required to move their eyes to the location of a stimulus that they have not seen in their blind field (Kentridge & Heywood, 2001). These findings demonstrate that both feature and spatial encoding in the absence of primary visual cortex (V1) can be preserved, although implicitly.

Hippocampal "Place" Cells

By far the most extensive animal literature concerning the representation of space has been focused on brain structures that determine how navigation is accomplished. Limbic structures such as the hippocampus and adjacent areas are necessary for long-term spatial memories that guide an animal to where it wants to go (Nadel, 1991). Many cells in the hippocampus respond when the animal is in a particular place in its environment (i.e., "place cells"). However, what is most important for the present purposes is that there are a subset of visually tuned cells in the hippocampus that respond to locations in the environment (Rolls et al., 1989). An animal may move to different locations within a room but the visually tuned cells will respond when a stimulus appears at a certain location, say a spot on the north side wall. They code space in environmental coordinates independent of body location or orientation.

Consistently, human ability to learn a visual-spatial maze in a paper-and-pencil task is disrupted by hippocampal damage, although this deficit is more likely to occur with right than left hemisphere damage (Milner, 1965). Right parietal damage in humans is also more likely to cause spatial maze learning deficits than left (De Renzi, Faglioni, & Villa, 1977; Newcombe & Russell, 1969; Ratcliff & Newcombe, 1973). In other words, right hemisphere damage to the hippocampus *or* the parietal lobe can produce spatial memory deficits that affect spatial navigation. Given the connections between the parietal lobe and hippocampus through the parahippocampal and entorhinal cortex (Suzuki & Amaral, 1994), this is not particularly surprising.

Precentral Sulcus of the Frontal Lobe

In addition to areas that are involved in establishing spatial maps that remain in long-term memory and are valuable for navigation, there are other areas that store spatial information for shorter time intervals. Patricia Goldman-Rakic and her colleagues demonstrated that neurons in the principal sulcus of the dorsolateral prefrontal cortex respond to the spatial location of a stimulus during a delay when the location is to be remembered (Funahashi, Bruce, & Goldman-Rakic, 1990, 1993). This area is topographically mapped with neurons in one cortical area responding to a particular region of space, and a lesion in this area will produce a blind spot in memory for stimuli presented in this region. Again, this area of the cortex is strongly connected to the parietal lobe (Cavada & Goldman-Rakic, 1989).

Spatial Coding and the Ventral Stream

One large part of the cortex that is often ignored in neurophysiological models of spatial processing is the ventral stream that supports object perception. An exception to this general rule is a model proposed by Desimone and Duncan (1995) in which object features at different locations compete for selection.

They point out that from primary visual cortex (V1) anteriorly through V2, V3, V4, and so forth, spatial topography tied to the retina changes from finer to courser grain. The more posterior neurons have smaller receptive fields, and the size of the receptive fields increases, going progressively more anterior within the temporal lobe (Gattass, Sousa, & Covey, 1985). However, neurons with very large receptive fields (areas TEO and TE in the most anterior part of the temporal cortex) still have a location within each receptive field where firing is more vigorous, with a gradual fall-off in firing rate from this point out. One interesting property of neurons in some areas within the ventral stream is that many of the neurons' responses can be changed by attention (Moran & Desimone, 1985). For instance, when two stimuli are presented within the receptive field of a V4 cell and one is to be ignored, modulation of the neuron's receptive field occurs, but when the ignored stimulus is placed the same distance away but outside the receptive field of the recorded cell, no modulation is observed. When the target and distractor are placed in the display such that they both project to locations that are within the boundaries of the receptive field, they compete for attention, but when they cross boundaries, competition is not required.

As a result of findings like these, Desimone and Duncan (1995) proposed a biased competition model in which attention is thought to be represented by competition within the local space of each neuron (i.e., competition for the receptive field of the neuron). There are also competitive mechanisms between neurons for the information that "wins" each neuron's receptive field response. When covert attention is allocated to a location outside the receptive field of a monkey V4 cell, the receptive field of a neuron at fixation can change in ways that elongate its field toward the covertly attended location (Connor, Gallants, Preddie, & Van Essen, 1996). The point of fixation and the point of attention interact to change the spatial field of neuronal firing.

This collection of evidence supports the claim that some types of space are represented in the temporal lobe. The patient work demonstrates that competition for space is limited without parietal input. Competition that changes neuronal space does not occur normally when bilateral parietal damage is present (Friedman-Hill, Robertson, Ungerleider, & Desimone,

2003). Balints patients are unable to explicitly report the relative location of even a simple item in a display, although they detect it. The evidence I discussed supporting implicit spatial representations in these patients may arise from spatial maps in the temporal lobe, although other candidates are clearly available, as is obvious from this overly simple and brief review.

Summary

In sum, there are various areas within the brain that represent space and could produce the implicit spatial effects found with parietal damage. RM could not be tested on maze learning or motor learning due to the severity of his spatial deficits. He could not reach or point in the correct direction of a simple light, let alone draw, trace, or navigate his environment. Perhaps other measures, such as functional imaging, would be useful in determining what areas of the brain are most active in representing the complex implicit spatial information that I have discussed in this chapter. Although there have been many imaging studies of spatial attention and spatial abilities, there has been no imaging evidence to my knowledge addressing questions of implicit spatial representations and the many candidate systems that might support them. Implicit spatial information could be represented in ocularmotor maps, but it is hard to conceive of these maps as the basis for processing global levels of patterns like those in Figure 3.20 that RM did not see, but that nevertheless influenced his performance. The implicit effects may also be due to bimodal cells of the premotor cortex or putamen that code space with reference to body parts, but recall that RM's explicit spatial awareness of his own body parts was not affected. Alternatively, the implicit effects might arise from coding in environmental space supported by limbic structures involved in memory. However, spatial memory problems were not observed with RM, and casual observation demonstrated that he was clearly able to learn new environments. He could tell us where to turn to go to the testing room, the men's room, or the elevator, and this occurred in several different laboratories where he was tested over the many years we studied his spatial abilities.

Alternatively, the implicit spatial effects may represent spatial maps of the ventral system that are involved in perceptual organization of visual input into objects with hierarchically arranged levels of object/space structure. Simultanagnosia could result from difficulty in switching between hierarchically arranged frames of reference.

Whatever the case turns out to be, there is clearly a great deal of spatial encoding that occurs without conscious awareness of that space. Space is normally difficult to ignore, but when the brain is damaged in particular ways, space may no longer be available to awareness. It is quite possible that parietal lobes may function to bring selected spatial representations into awareness by controlling the spatial system that is most relevant for the task at hand (Gross & Graziano, 1995). Another possibility is that parietal functions act to integrate the various spatial maps found throughout the brain into a master map of locations (see Treisman & Gelade, 1980) that then supports the experience of a unified space (see Andersen, 1995, and Mesulam, 1999, for models that are consistent with this approach).

6

CHAPTER

Space and Feature Binding

In previous chapters I argued that the space we perceive is represented by multiple spatial reference frames. For the most part, these spatial frames represent the structure of shapes or objects, their spatial relationships to each other, as well as the planes on which objects appear. Objects, groups of objects or whole scenes require some type of spatial skeleton in order to take form. This skeleton includes several spatial frames, creating a hierarchy of reference frames, somewhat akin to that proposed some years ago by Palmer (1989). The major focus in this chapter concerns how features that are not easily defined by spatial frames are integrated within object boundaries (e.g., the green of a green ball).

There are many visual features that need not conform to spatial coordinates (e.g., color, texture, brightness, etc.) yet are properties of objects. For instance, a ball may be green or red, rough or smooth, appear in bright light or in a dark corner, and still be seen as a sphere. We might speak about green as being part of the ball, but it is quite a different part than the ball's curved boundary. The only apparent spatial restriction on the color green is that of the edge of the ball itself. Although this is our perceptual experience, it does not appear to be an accurate portrayal of processing that takes place before "a green ball" is perceived. In the initial few milliseconds before perceptual awareness, many of the features we encode seem to be unconstrained by spatial boundaries. Green can be detected without the need to locate or quantify it, leaving it susceptible to miscombinations with other features that are spatially specified (e.g., shape).

As discussed in previous chapters, neurobiological evidence has shown that features such as color, form, motion, and shape activate different specialized areas within the primate cortex (Livingstone & Hubel, 1988; Moutoussis & Zeki, 1997; Zeki, 1978). Behavioral data have supported this independence. For instance, color and shape can be detected independently and misconjoined in perception to form an erroneously colored "object." For instance, when shown a brief presentation of a red A and a blue B, participants might be quite confident they saw a red B and a blue A.

These errors were first demonstrated by Treisman and Schmidt (1982) who coined the term "illusory conjunctions" (ICs) to describe the phenomenon. Treisman (1996) proposed that features (A, B, red, blue) are coded in separate "feature maps" with A and B represented in a map of letters or shapes and red and blue represented in a map of colors. In order to know which colors go with which shapes, another process must occur that binds the appropriate color and shape together. She proposed that this process is spatial attention. Attending to the location of the A conjoins the A with the color in the attended location (in this case red). It follows that if attention to the location of the A can be disrupted in some way, then the A could be perceived as the incorrect blue. Either red or blue would be conjoined with the A if spatial attention could not be engaged or if the spatial map on which attention relies disappeared, for instance through brain damage.

The way in which Treisman and Schmidt (1982) decreased the ability of normal perceivers to direct attention to the location of a shape was to use measures that directed attention elsewhere, showing the display for a very brief period of time, and in some cases masking the stimulus. She found that under these conditions, illusory conjunctions appeared regularly for normal perceivers. When attentional allocation to a location of an item was disrupted, accurate binding was also disrupted. It was not as if shape and color simply fell apart and were perceived as different properties unconjoined at all (although this can happen too; see Ashby, Prinzmetal, Ivry, & Maddox, 1996). Rather, shape and color were bound incorrectly.

Attentional involvement in conjoining features has received additional support from findings demonstrating that searching for a conjunction (say, a red A) among distractors with features in common (e.g., red Bs and blue As) requires a serial search through the display. As the number of distractors increases, the time to find the target increases (Treisman & Gelade, 1980). This time can be decreased by grouping and other perceptual organizing principles that affect the manner in which the system rejects distractors, which can then guide search more efficiently (Wolfe, 1994). But for the most part, a serial scan is required to find a conjunction target in a cluttered display (Treisman, 1988). The time to find a distinct feature (e.g., a red A among blue Bs) is not affected by the number of distractors in a display, consistent with parallel search. Both red and A pop out from the background.

One way to eliminate serial search patterns for conjunction displays is to cue the location where a target is likely to appear (Treisman & Gelade, 1980). When normal perceivers are given information allowing them to direct attention to the location of an upcoming target, reaction time to detect the target is as rapid as when the target is presented alone. However, if the target is presented where it is not expected and in a cluttered array of distractors with features in common, evidence for a serial search again appears.

These three bits of evidence together (ICs under divided attention conditions, increased time to find a conjunction but not a feature target as the number of distractors increases, and eliminating distractor effects in conjunction search by spatial cuing) represent converging support for the special role of spatial attention in feature binding. They provide the cornerstones of feature integration theory (FIT) proposed by Treisman and Gelade in 1980, which has had substantial influence in the cognitive sciences, vision sciences, and neurosciences.

There have been several alternative explanations of one or another of these cornerstones over the years (see Chelazzi, 1999; Duncan & Humphreys, 1989; He & Nakayama, 1992; Nakayama & Silverman, 1986, for examples of some controversies), but few have attempted to account for all three of the major phenomenon that support FIT (although see Luck, Massimo, McDermott, & Ford, 1997). For example, the biased-competition model (BCM) proposed by Desimone and Duncan (1995) claims that features compete for processing within defined areas of the visual field in parallel. Through competition and inhibition, the correct combination of features surfaces without the need for attentional search. A bias toward one feature or another can occur through top-down information such as the designation of a target form or color (e.g., the instruction to look for the red A) and/or bottom-up mechanisms that determine perceptual saliency such as luminance contrast, feature similarity, and other parameters that affect salience. BCM does a good job of providing an alternative theory to account for differences in conjunction and feature search performance. It is not my goal here to resolve the different theoretical approaches (instead see Treisman, 1999), but it is worth emphasizing that the BCM was proposed to account for visual search and spatial cuing results. Its ability to explain the phenomenon of illusory conjunctions is less clear.

I will touch on these alternatives again while discussing recent findings that shed new light on the "binding problem" (see Robertson, 2003), a problem that has puzzled scientists since the relatively modular architecture of the cortex was discovered. It took the emergence of cognitive neuroscience for the field to realize the wealth of behavioral data that had already been collected to test FIT that might then be applied to issues of binding within brain sciences. I will start this chapter with evidence from neuropsychology that clearly demonstrates the reality of a binding problem in everyday life when damage to particular areas of the cortex, specifically the parietal lobes, is present.

□ The Effect of Occipital-Parietal Lesions on Binding

Illusory Conjunctions

The first indication that lesions of the human parietal lobes might disrupt feature integration came from the work of Cohen and Rafal (1991). They tested a patient with unilateral right extinction who was biased to attend to the ipsilesional side of space after a left hemisphere stroke affecting posterior areas. Under conditions of brief stimulus exposure, more ICs were found when stimuli were presented on the right (extinguished) side than when they were presented on the left (attended) side. Although these findings were provocative, they were not conclusive because they could be explained by a variant of the methods that produce ICs in normal perceivers (i.e., when attention is diverted and stimuli are briefly presented). If normal perceivers were encouraged to attend to the left, more ICs would be expected for stimuli that appeared on the right, as was the case when attending to the left was produced by an act of nature.

Later, my colleagues and I reported that patient RM with Balints syndrome and bilateral occipital-parietal damage (see Figure 5.2) produced a high rate of ICs (up to 38% in early testing) even under free viewing conditions. The need to divert attention to one side or the other combined with brief stimulus presentation was unnecessary.

Given the severity of RM's spatial deficits (as described in chapter 5), it was not surprising that his ability to attend to locations in space was nearly completely lost (see Ungerleider & Mishkin, 1982). Even when cued, he was initially at chance in reporting where the cue or a subsequent target occurred, and he was no better than chance at reporting whether two sequentially presented stimuli were in the same or different locations. In other words, even under conditions where attention is normally automatically drawn to a location, RM was not aware of the location of a stimulus nor was he aware that two sequentially presented stimuli were in the same or different locations (Friedman-Hill et al., 1995). He was able to perceive the two stimuli presented separately in time, but he did not know their locations. In addition, sequential presentations did not produce ICs. However, for simultaneous presentation, ICs appeared even when the stimuli were simple (two letters in two different colors) and shown for up to 10 seconds (Friedman-Hill, et al., 1995; Robertson et al., 1997). Since RM only saw one item at a time (simultanagnosia), we asked him to tell us what letter he saw on each trial and its color *as it appeared to him*. ICs were prevalent over many different testing sessions and exposure durations.

Similar IC rates were observed whether the stimuli were presented for 500 ms or for 10 seconds. This too would be expected if a person lost external spatial maps as a result of brain injury. Without a spatial representation, attentional allocation to a location would be impossible even if objects remained on the screen. The important point of this discussion is that when explicit spatial knowledge of the external world was unavailable, binding surface features together was affected. RM made very few guessing errors. He reported seeing a color or letter that was not presented on very few trials. Overwhelmingly, his errors were conjunction errors.

When probed about his perceptual experience while performing the task, RM told us that he was reporting the letters as he saw them. He commented with statements like "When I first look at it [the letter], it looks green and it changes real quick to red," the letters on that trial being red and green, or "I see both colors coming together." The colors and letters seemed to randomly join together in his experience. But did he explicitly see all the colors in a stimulus, and if so, did the most salient colors emerge in awareness, as would be expected by accounts based on bottom-up saliency?

In another study, we asked RM to name all the colors that he saw in displays that contained either two or three colors and found that he did know explicitly what colors were in each display (Figure 6.1). However, the amount of any one color on the screen did not seem to matter. Marcia

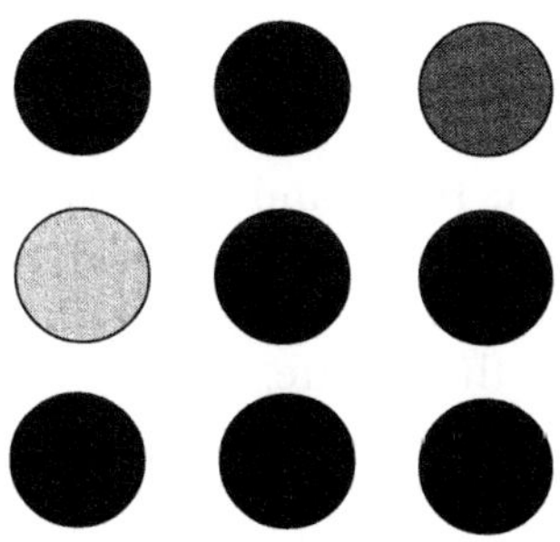

FIGURE 6.1. An example of a stimulus presented to RM by Marcia Grabowecky with three different colored circles, one red (light grey), one blue (medium grey), and seven green (black). RM reported seeing only one of each color.

Grabowecky presented him with a cluster of colored circles in which one or two were one or two colors and the rest were a third color (e.g., one red, one blue, and seven green). When asked to report the colors and number of circles he saw in each color, he consistently reported one of each (e.g., one red, one blue, and one green). He seldom deviated from this pattern, saying that he saw only one of each color that was actually in the display. When asked if he saw more than one circle in any color or more of one color than another, he said no. These results would be expected on the basis of FIT. Without individuating the circles, the colors were bound to the space of the circle he did see. The colors were either rapidly interchanged on the one shape over time or, less intuitively, were present in parallel within the same circular form (Robertson, 2003). Each color was represented with the shape of the circle even though one color (in this example, green) should have been the most salient color in the display. The colors were registered, but without explicit space, their distributions were not.

We also demonstrated that ICs in free view were not limited to color and shapes. They occurred between shape and size (Friedman-Hill et al., 1995) and shape and motion (Bernstein & Robertson, 1998) as well. Testing IC rates between orientation and color was problematic because RM was very poor at reporting orientation even when only one item was present in a display. Given that spatial orientation is also a spatial property defined by relative orientation, it is not surprising that the orientation of the objects RM did see were explicitly unknown to him (Robertson et al., 1997). He was also unable to judge the sense of direction of a familiar form (e.g., whether the letter F was normal or reflected) or the size of the shapes he saw. The fundamental properties that define a spatial reference frame (as described in chapter 2) were unavailable. Without a spatial reference frame on which to hang the features, ICs were evident even in paper-and-pencil tests as well as in everyday life (e.g., on one occasion he reported that a house appeared to move when a car was going down the street).

Elevated IC rates have now been verified in at least two additional Balints patients, one tested by Hanaff, Michel, and myself at the INSERM in Lyon, France (unpublished data, 1996) and one tested by Humphreys and his colleagues in England (2000). Humphreys et al. (2000) also demonstrated that the relative size of items in the display and contrast polarity (black and white) did not affect the IC rate, nor did connecting items by a line that would make it more likely that the patient would see the two items as a whole (Figure 6.2). Gestalt principles of grouping clearly affected what shapes were explicitly seen (i.e., binding elements on the basis of lines, angles, collinearity, etc.), but they did not affect binding of surface features such as color and shape or shape and polarity. Binding parts into

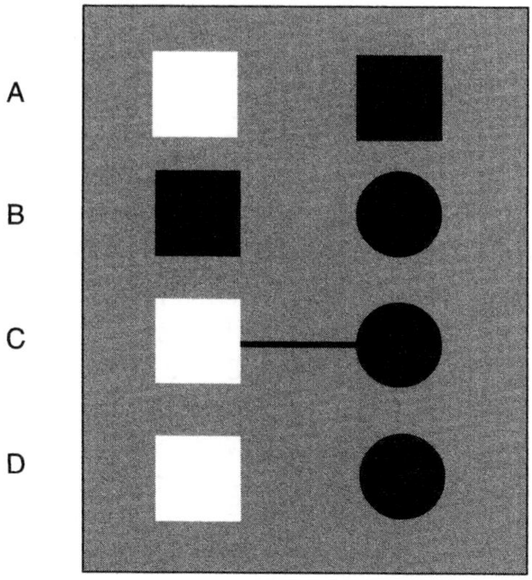

FIGURE 6.2. Example of stimuli used by Humphreys et al. in a study with patient GK. (a) can be grouped by shape, (b) by contrast, (c) by connectedness, and (d) is not grouped. (Adapted from Humphreys et al., 2000.)

at least one object was preserved, consistent with the clinical observation of simultanagnosia, but binding surface features to shape was deficient.

The evidence concerning deficits in binding demonstrates that bilateral occipital-parietal damage produces a real-life binding problem in addition to the spatial deficits that have long been observed. Grouping and binding features such as lines and angles into an individual object appear *relatively* intact in Balints patients and can sometimes be affected by manipulations that affect perceptual organization in normal perceivers. However, individuating one object from another, attending to the location of an object, and binding properties of objects accurately do require intact parietal lobes. It appears that some type of spatial signal from the parietal lobe guides attention and interacts with ventral areas that encode different features in specialized cortical areas (Robertson, 2003). This signal might synchronize firing between neurons in separate feature maps (Koch & Crick, 1994; Singer & Gray, 1995), inhibit firing to irrelevant features within specialized neurons (Friedman-Hill et al., 2003; Desimone & Duncan, 1995), co-locate activity along different dimensions (Garson, 2001), or co-locate activity in preattentively encoded but separate feature maps (Treisman, 1988). Whatever the ultimate explanation, the data from patients demonstrate that binding of surface features that are represented relatively separately in the ventral pathway is facilitated by spatial attentional functions of the parietal lobes. They further suggest that explicit spatial awareness is necessary for proper binding of surface features.

Visual Search

If the IC rates observed in cases of Balints syndrome are due to spatial deficits, as FIT predicts, then these patients should also have great difficulty serially searching for a conjunction in a cluttered array but have little, if any, difficulty searching for a unique feature. According to FIT, spatial attention is not required to determine whether a particular feature is present or absent in a stimulus, but when features must be combined, a serial search may be necessary to co-locate the features that form the conjunction target. Given RM's high IC rate and severe spatial problems, we predicted that he would be very poor at searching for conjunctions in visual search displays but good at detecting single features.

These predictions were confirmed. We first casually presented RM with between 20- 40-item search displays on pieces of paper placed in front of him (Figure 3.1). When he was asked to report whether a red dot among yellow and blue distractors was present (feature search), he was able to do so accurately, although he could not report the target's location. However, he was unable to find the conjunction of a red circle with a line through it in a conjunction display even after viewing the display for 30 seconds or more. He would essentially become glued to one item in the display and could not move his attention elsewhere. Since we were unable to obtain reasonable data for conjunction search with these displays, we changed the stimuli to make the task trivially easy (at least for normal perceivers).

Display sizes of two, four, or six items were presented on a computer screen (Figure 6.3) for up to 10 seconds and both errors and reaction time for RM to verbally report whether or not the target was present were recorded. For both conjunction and feature search the target was a red X. In each conjunction search display the distractors were red Os and green Xs. In each feature search display the distractors were all green Xs or all red Os. Each item was a salient, filled-in letter (1°) and the displays subtended a 10° x 10° area.

Even in a conjunction display, normal perceivers would find this an easy task, given the sparsely located and small number of items in each display. But RM found it very difficult and made many errors. His mean response times for correct trials were between 2 and 4 seconds on average and were so variable that they were rather meaningless, but his pattern of errors were quite informative. When the target was present (Figure 6.3a), he missed it only 4% of the time, but when it was absent (Figure 6.3b) he confidently reported its presence 38% of the time. This would be the expected outcome if the color of the distractor O (red) had been miscombined with one of the distractor Xs (green). In this case the target would be absent in the display, but when he saw an X, it could appear to him as either green or red, resulting in a large false alarm rate.

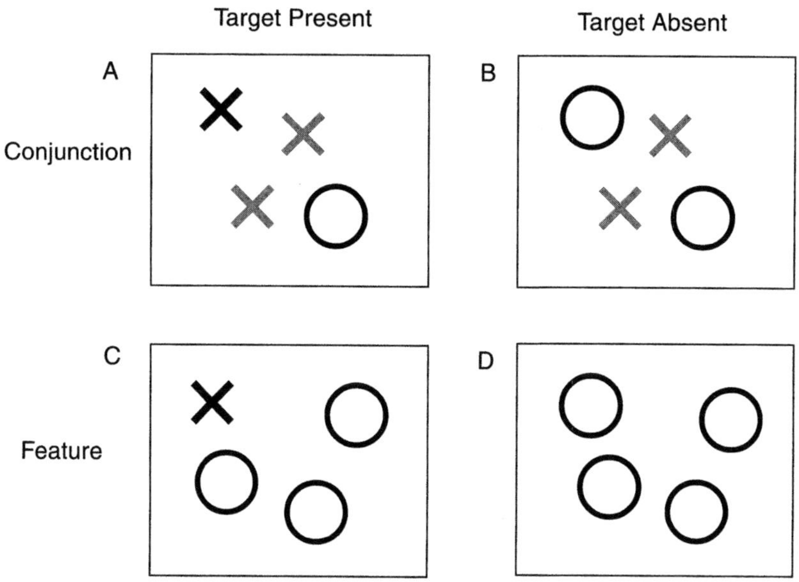

FIGURE 6.3. Example of stimuli used to test RM's visual search abilities. The target (a red X, represented by black) was present (a) or absent (b) among distractors that required a conjunction search, or it was present (c) or absent (d) among distractors that required only the detection of the feature X or red.

Performance was completely different for feature search (Figure 6.3c and 6.3d). (According to FIT, when the distractors were all green Xs, the presence of a unique color should have been sufficient to respond yes, and when they were all red Os, the presence of a unique letter should have been sufficient to respond yes.) As expected, RM was relatively good at detecting features despite his simultanagnosia. Although he did make about 4.5% errors, there were no differences between the number of misses and false alarms. Furthermore, reaction times based on correct trials did not increase linearly over display size as would be expected if he serially searched for the target from item to item. In fact, the slopes were somewhat negative as distractors increased. Although slower than for normal perceivers, the features "popped out" for him. The critical difference was that he did not know the location of the features he saw.

In another experiment we confirmed RM's difficulty conjoining features when he was instructed to select according to color and report the letter. We asked him to simply report the letter that was red in a two-item display. Again, he was very poor at this task. He made between 30% and 38% errors in different blocks of trials. He showed no obvious sign that he was aware of making this many errors, which would be consistent with his actually seeing the letter he reported as red. Again, he did not know the location of the items he reported.

It is very difficult for normal perceivers to conceive of a world in which color and form would not be located somewhere, but that would be our perceptual experience if explicit spatial awareness were lost. It is easy to imagine color or form being in the wrong place, but what would the world look like if features had no place at all, just detectors informing us of their presence? If all we were left with were detectors for common shapes and basic features without locations on which to hang them, all we would have left are of the features themselves. Under such conditions, features could easily be erroneously combined, and they are.

☐ Additional Evidence for Parietal Involvement in Feature Binding

Findings from studies using functional imaging, evoked potentials, and transcranial magnetic stimulation (TMS) have all supported a role for the parietal lobes in conjunction search. TMS (Figure 6.4) essentially produces a very brief disruption of function (in the millisecond range) in normal brain tissue in a targeted area. When Ashbridge, Walsh, and Cowey (1997, 1998) used this procedure with normal perceivers, they found that pulses applied on the scalp over the parietal lobe slowed conjunction search but did not affect feature search. Consistently, Corbetta, Shulman, Miezin, and Petersen (1995) found parietal activity in a PET study during conjunction search but not during feature search (Figure 6.5). Importantly, posterior temporal areas were activated in both cases. Activation of these areas would be expected in both conditions because features should be registered by their respective visual areas whether they are present in a conjunction or feature display. Studies measuring electrical activity from scalp electrodes have also reported a dissociation between feature and conjunction search, with one component of the ERP wave (N1) producing the same response for feature and conjunction searches but an additional component (P1) affected only by conjunction search (Luck & Hillyard, 1995).

Both behavioral and neurobiological evidence support feature and conjunction search as qualitatively different. The neurobiological evidence demonstrates that conjunctions do not simply increase neural activity within given areas of cortex, but rather engage additional areas outside those involved in basic feature registration. These outside areas are in parietal lobes and therefore part of the dorsal "where" pathway that encodes spatial information explicitly (see chapter 5). Converging evidence reveals a pattern of interaction between feature representations in the ventral pathway and spatial functions of the dorsal pathway that appear necessary for accurate feature binding and serial attentional search.

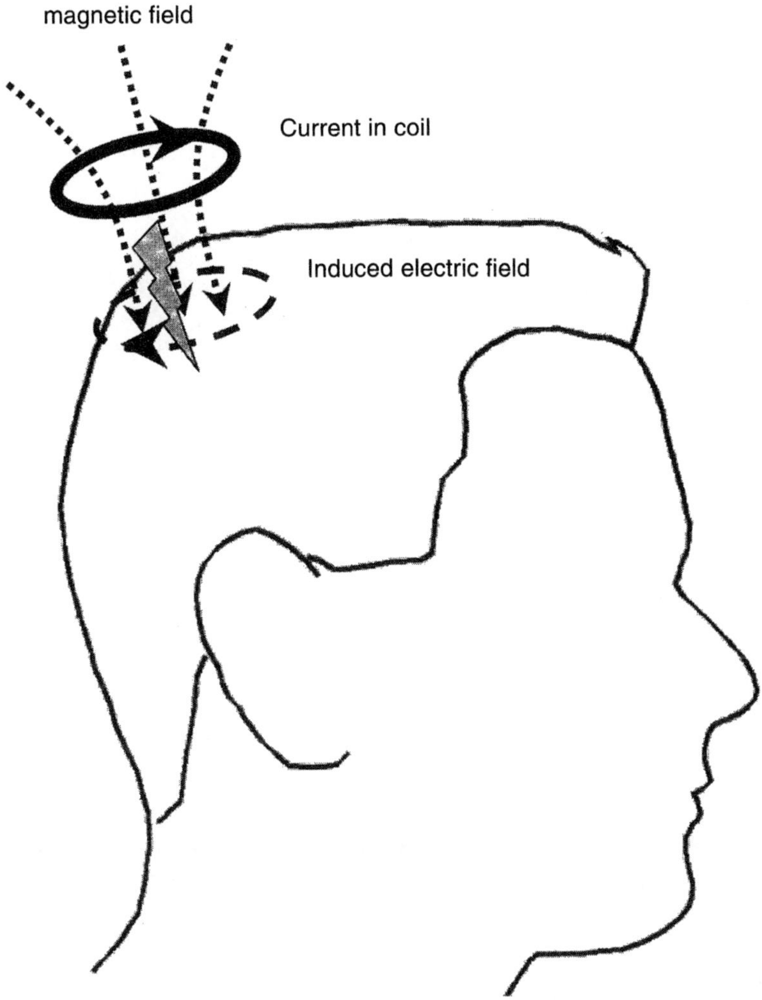

FIGURE 6.4. Trancranial magnetic stimulation (TMS) is a procedure in which a magnetic pulse is generated by a coil placed on the scalp that produces a very brief disruption of function for a few milliseconds. The area of the brain that is disrupted is about 1 cm depending on the type of coil and amplitude of the pulse signal.

There are strong connections between the parietal and temporal lobes, most notably between LIP and temporal areas V4, TE, and TEO (Baizev, Ungerleider, & Desimone, 1991; Blatt, Andersen, & Stoner, 1990), see Figure 6.6. The data discussed in the previous section may reflect a functional role for the posterior connections between dorsal and ventral areas that in part reflects binding surface features.

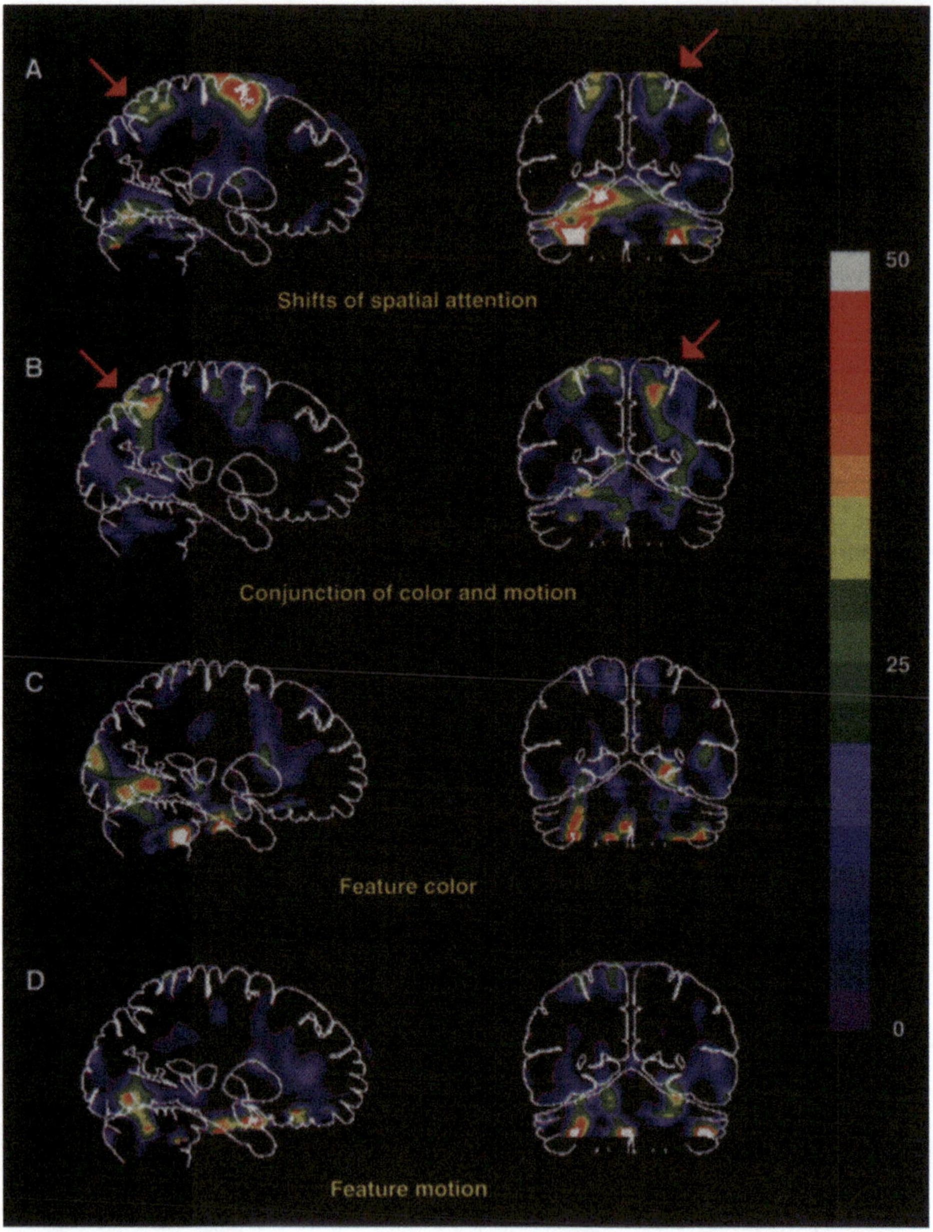

FIGURE 6.5. Brain activity during conjunction and feature search in a PET study. Note the ventral activity for both types of search with the addition of parietal activity for conjunction search. (Reprinted with permission from Corbetta, M., Shulman, G., Miezin, F., & Petersen, S., Superior parietal cortex activation during spatial attention shifts and visual feature conjunction. *Science, 270,* 802–805. Copyright © 1995 American Association for the Advancement of Science.)

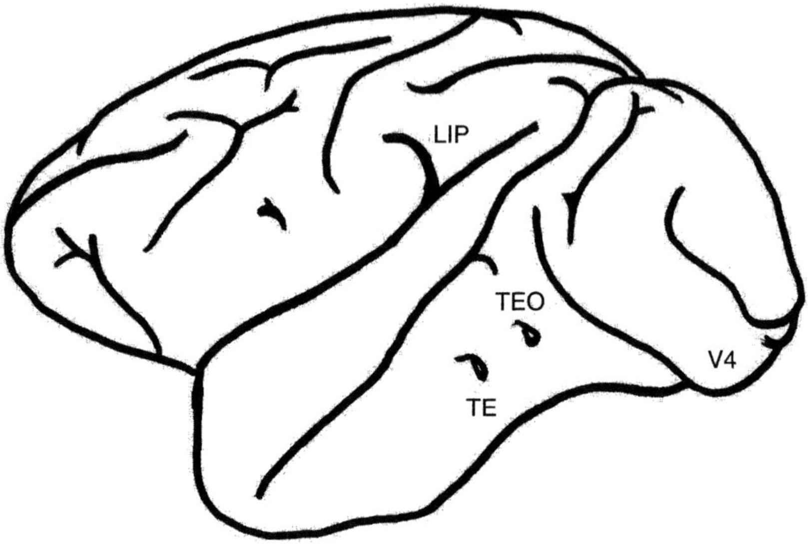

FIGURE 6.6. Cartoon showing general areas of V4, TE, TEO, and LIP.

The issue of exactly what functions the parietal lobes play in this type of binding has been a matter of some debate. For instance, when comparing feature versus conjunction search performance, issues of difficulty and saliency often arise. Parietal activity may be more pronounced under conditions of difficult search or lowered visual saliency. But note that both of these conditions would call for additional attentional resources. Others have suggested that parietal activation in conjunction search reflects adjustments to the size or movement of a spatial window (see Ashbridge, Cowey, & Wade, 1999). Still others argue that increased parietal involvement reflects an increase in attentional inhibition (Chelazzi, 1999).

These are all potential candidates when considering both biological and cognitive evidence for differences between conjunction and feature search performance. The one case in which difficulty, resources, or inhibition lack explanatory power is that of illusory conjunctions. Reporting one of two letters on a computer screen and its color is not a difficult task. Yet, parietal lobe damage consistently both increases IC rates and disrupts conjunction search performance (see Arguin, Cavanagh, & Joanette, 1994; Cohen & Rafal, 1991; Eglin et al., 1989; Esterman et al., 2000; Pavlovskaya, Ring, Groswasser, & Hochstein, 2002, for examples in unilateral cases).

The parietal lobes are part of a large network that involves spatial representation, spatial attention, feature registration, feature integration, and more. In fact, the binding deficits may well reflect a distributed network with connections between selected parietal areas and both cortical and subcortical areas. A recent intriguing report of increased ICs and decreased spatial abilities involved a patient with unilateral pulvinar damage without neglect or extinction (Ward, Danziger, Owens, & Rafal, 2002). When color/letter conjunction errors were compared for contralesional and ipsilesional displays, there were no differences in feature errors (about 2.5%), but ICs were 19% for contralesional displays and only 2% for ipsilesional displays.

These findings are consistent with suggestions that binding surface features (or what Treisman (1996)calls property features) might be facilitated by reentrant pathways into primary visual cortex through the pulvinar (Treisman, 1998). Neither she nor I are advocating that binding takes place in the parietal lobe per se. Rather, it occurs as part of a system of spatial representations that guides attention to spatial locations within selected frames of reference. The inferior parietal lobe is a critical part of this network, one that is necessary for explicit spatial representations of the outside world.

☐ Implicit and Explicit Spaces and Binding

The discussion in the previous section leads naturally to the question of whether features are bound implicitly or require explicit spatial processing (see chapter 5). As noted in chapter 5, indirect methods have shown a multitude of implicit effects in perception, including implicit spatial information for patient RM. The evidence I have discussed so far demonstrates that some sort of spatial medium is necessary for accurate binding of features such as color and shape. Co-locating features seems to be a key ingredient.

Of course, some sort of spatial medium is also necessary to bind the parts of objects together to form a single object for normal perceivers as well as for patients with simultanagnosia. The evidence for implicit space collected with RM (see chapter 5) is therefore a relief in many ways because it allows for at least one solution for how patients like him might be able to see an object, albeit only one at a time. Lines and angles and other features that define the shape of objects may utilize spatial maps within ventral pathways that need not reach spatial awareness. Binding features together to form the shape of an object requires connecting, say, an angle to a curve or grouping unconnected elements guided by such principles of organization as common motion or proximity. The junction that connects the two lines that form a T must be encoded as in the same location in order to be seen as the letter T, but this is a different computational problem than binding the color green to the letter T. Likewise, the elements of a group of moving dots contain spatial relationships that are maintained over spatial transformation independent of their color. There is a frame that holds the dots together as a group or object in perception. In either case, once this type of binding is complete, the bound items themselves must then be individuated from other bound items. Common to both is the need for a spatial reference frame to bind the parts of objects into wholes. The fact that patients with severe spatial deficits continue to perceive grouped objects suggests that at least some spatial maps remain functional below the level of spatial awareness.

Tests of implicit space with RM (discussed in chapter 5) have only included 2-D space to date, but they demonstrate that spatial relationships between unconscious representations of items can be implicitly encoded. As discussed previously, the spatial relationship between the location of individual features in a display and a central target were encoded, as were the spatial relationships between global and local elements in a very different type of stimulus (e.g., Egly et al., 1995). This was apparent even though RM did not explicitly know the location of the elements and could not report more than one at any given moment. These effects bring us back to the question of whether binding of surface features can also occur without spatial awareness (i.e., implicitly).

There is some evidence that binding features such as color and form can happen implicitly, but the literature is not conclusive on this point. One bit of evidence that is sometimes referenced to support binding color and form without attention comes from negative priming studies, but here, too, attention seems to be required. This procedure was introduced in the cognitive literature some years ago, showing that a distractor on one trial slows response time on the next when it becomes the target. For instance, when normal perceivers are asked to report the letter in green on each trial, a distractor form (say, red x) on one trial slows response time on a later trial when the x (now green) becomes the target (Tipper, 1985). In fact, DeSchepper & Treisman (1996) showed that negative priming could last for days or even months under the right conditions, demonstrating that the integration of color and form can be retained in memory over surprisingly long durations (although see Wolfe, 1998). If only the form is retained in memory, positive priming occurs.

But the question of importance for the present discussion is: Must a distractor such as a red x on the prime trial in a negative priming study reach conscious awareness before inhibition occurs? It seems possible that attention directed to the to-be-ignored item during Trial 1 might be required, and it could be at this stage that binding of the letter and color in the distractor happens. Evidence reported by Fuentes and Humphreys (1996) seems to support this idea. They tested a patient with left visual extinction due to right hemisphere stroke on a matching task with sequentially, centrally presented letters. These letters were always blue.

During prime presentation, irrelevant flankers were placed to the left or right of the central blue letters. The flanker letters were green. When the probe was a letter that had been an irrelevant green flanker appearing on the extinguished side in the prime display, positive rather than negative priming occurred (e.g., an x, whether green or not, facilitated response time). But when the probe was a letter that had been a green flanker on the nonextinguished side in the prime display (where attention would be relatively normal), the usual negative priming effects were found.

These results demonstrate that at least in patients with unilateral extinction, attention is required for conjunctions to be encoded into memory even if those conjunctions are to be ignored. But what happens to these conjunctions after they are ignored? There is other evidence that they are not easily recalled even directly after stimulus offset. DeSchepper and Treisman (1996) asked their normal perceiving participants to pick out the shape they had just seen from several alternatives, all being presented immediately after a subset of trials, but found that recall was typically poor. In another study the size of the distractor on the prime trial was changed when it became the target on the probe trial, and rather than negative priming, positive priming appeared (see Treisman, 1998). When a feature was changed, the inhibition was no longer present, suggesting that the representation that produces negative priming is tightly bound in memory.

When these findings are interpreted in light of those reported by Fuentes and Humphreys (1996), it seems that inhibition of objects happens almost immediately and is quite specific to the conjoined features. The evidence as a whole suggests that attending to conjunctions (and hence binding) is necessary for negative priming to occur. Although attended, inhibited conjunctions are very soon forgotten in explicit memory, but continue to be stored in implicit memory and influence the speed of a later response. As a result, data from negative priming studies do not support binding without explicit spatial attention. But is there other good support for implicit binding in the literature?

Perhaps the most problematic evidence against the claim that explicit binding of color and form only occur with spatial awareness were collected with RM himself. Wojciulik and Kanwisher (1998) used a very creative variant of the color/word Stroop task to examine this issue and concluded that preattentive binding does take place. Because these results are some of the most convincing evidence for preattentive binding to date, I will discuss them in some detail and then explain why I think they are inconclusive.

Wojciulik and Kanwisher (1998) created lists of 48 trials using four words and four colors. The words were either color words (green, yellow, brown, purple) or neutral words (short, ready, useful, careful) printed in green, yellow, brown, or purple. In the Stroop blocks, the "ink" colors the words were printed in were either incongruent or congruent with the word meanings (e.g., GREEN printed in yellow (GREENy) vs. GREEN printed in green (GREENg), respectively). Normal perceivers are faster when naming the ink color when the word is congruent than when it is incongruent (Stroop, 1935). The variant that was introduced when testing RM was the addition of a second word that was achromatic (a) and was one of the four words used in that block of trials (see Figure 6.7). The two words could either be both congruent with the color (GREENg, GREENa) or both be incongruent (YELLOWg, YELLOWa), and the distractor word could be either incongruent with the ink color (GREENg, YELLOWa) or congruent with the color (YELLOWg, GREENa). RM's task was to name the ink color in the display on each trial as rapidly as possible, ignoring both words. Note that the Distractor Incongruent (DI) and Distractor Congruent (DC) conditions contained the same features but they were combined differently. In the example in Figure 6.7 the features would be the words green and yellow and the colors green and no color. If RM implicitly bound features together, the reasoning was that the color green would be harder to report in the DC than in the DI condition because in the former the word YELLOW would produce interference when reporting the color green. But the question was whether they were bound implicitly in vision.

Another critical part of the experiment was an explicit binding condition in which RM was instructed to name a neutral word that was colored on each trial. For instance, if the pair of words were CAREFULg and USEFULa, then the correct response would be "careful". The displays were time limited, as his ability to report both words had increased to about

	Both congruent (BC)	Distractor incongruent (DI)	Distractor congruent (DC)	Both incongruent (BI)
	GREEN GREEN	GREEN YELLOW	YELLOW GREEN	YELLOW YELLOW
RT_1	621	671	836	975
RT_2	—	1173	1207	—

FIGURE 6.7. Example of colored Stroop stimulus pairs used by Wojciulik and Kanwisher (1998) to examine implicit binding in RM. Gray represents the color green (the top word of the pairs shown) and black represents a monochromatic stimulus (the bottom word of the pairs shown). (See text for details.)

73% in free viewing by then. To overcome his better than chance performance in free view, Wojciulik and Kanwisher (1998) varied the display times of the neutral words to obtain chance performance in selecting the colored word from the pair. This was then recorded as the display time at which he experienced explicit binding problems. Once chance performance on naming the colored word was obtained, that time was then used to present the experimental blocks in which RM was asked to name the ink color as rapidly as he could while ignoring the color words (average presentation time was 159 ms).

The results are shown under the displays in Figure 6.7. At the same display times in which he was unable to select the colored neutral word above chance, he was 165 ms faster to name the ink colors in the DI than the DC condition. Although the same features were present in both, the indirect measures suggested that RM's visual system did encode the proper feature combination.

There was some concern that these results may have been affected by differences in luminance between the achromatic value and the ink color used in the stimulus displays. Since we do not know what these stimuli looked like to RM, it was possible that the greater saliency of the achromatic word in the first experiment attracted attention and interacted with the color in the display. To examine this possibility, the investigators later tested RM again with the colored word being less saturated and with less contrast of the achromatic words. The words in each pair were roughly matched for brightness via experimenter observation. Although the difference in reaction time between DI and DC conditions was reduced to only 34 ms (a small difference when testing patients) and did not reach significant levels as they did before, the mean responses were still in the same direction in 9 of the 11 blocks that RM completed (significant by a sign test). The effects were not as robust but the trend was present, so brightness differences between the colored and noncolored words may not account for the entire difference.

If these findings do indeed signal preattentive binding, they are surprising for several reasons. First, if color and form are implicitly bound, why does RM experience illusory conjunctions ever? In fact, why do normal perceivers experience illusory conjunctions under impoverished conditions? Once attention is directed to a shape, its color should automatically arrive with it if binding has already been accomplished. Second, there is other evidence with normal perceivers that implicit binding does not occur. Lavie (1997) used a flanker task in which features of the target (color and letter) were either conjoined or separated and presented to the right and left of a colored target letter. This manipulation made no difference at all on responding to the target when attention was focused on the center letter. Conjunctions and features interfered equally. However, when attention was widened to include the location of the flankers, conjunctions then produced more interference than single features. If conjunctions were bound without attention, the size of the attentional window should have made little difference.

So why did RM show interference in the Stroop tests of Wojciulik and Kanwisher (1998) if spatial awareness is required to bind surface features together? There are a number of possibilities. One is that all combinations of features are bound together preattentively (Treisman, in press), and that without parietal lobes and their spatial functions, the wrong conjunctions can be selected for awareness. Another possibility may be tied to the fact that primary features (even those like color and form that are registered in specialized cortical areas) are bound in early subcortical and cortical areas (e.g., V1, V2). There are spatially isomorphic maps in early vision where features are represented in the same location or very nearly in the same location. However, very soon, surface features are transferred to relatively specialized areas, and the question then becomes what mechanisms bring them back together. Space in very early visual areas could still contribute to performance but may be so weakly represented at the time of perceptual awareness (supplanted by other frames of reference) that they are basically overwritten. But in RM, whose perception is impoverished spatially, the effect of primary vision perhaps could continue to be observed.

Other, perhaps less interesting, possibilities are that the methods used by Wojciulik and Kanwisher (1998) were somehow inadequate (e.g., the control task was more difficult than the experimental task, brighter words attracted attention, the control task contained different words than the experimental task, etc.). To me it seems that the major question concerns what the stimulus looked like to RM, for his perception is not at all like what we experience. Could his abnormal explicit perceptions have produced the effects? We simply do not know.

It is difficult to explain how the data that offer some support for implicit binding with RM are related to his explicit binding problems. It is also unclear how brightness differences would interact with binding. Until evidence for implicit binding with normal perceivers is found, it will remain tentative whether or not surface binding occurs implicitly under normal conditions. Given the evidence as a whole, perhaps the most parsimonious conclusion at the present time is that the type of binding that requires information encoded in different specialized areas of the brain requires an explicit representation of space to be explicitly bound in awareness.

This conclusion should not be construed as applying to all types of binding. Binding as a general process almost certainly occurs at different levels of processing (see Humphreys, 2001). As pointed out previously, binding lines and angles together or grouping features to form objects does seem to occur in spaces that are represented below the level of spatial awareness. One need not obtain explicit spatial awareness to perceive objects.

Before leaving this topic, I should mention that another type of binding that presumably happens late in processing also seems to require explicit spatial awareness. For instance, automatic eye blink responses to looming objects are absent in Balints patients (see Rafal, 1997). Thus, binding the perceived object to action also seems disrupted without spatial reference frames of the external world.

☐ Summary

Spatial attention is clearly important in correctly binding features such as color, size, and motion to shape. When explicit spatial knowledge is severely affected, directing attention to a location is compromised, and binding deficits for surface features appear even under free viewing conditions (Prinzmetal, Presti, & Posner, 1986; Prinzmetal, Diedrichson, & Ivry, 2001). Detection of these features is relatively unaffected in such cases, but while unique features can be detected, their locations are unknown. These findings support fundamental proposals of FIT; features can be detected without spatial attention, while conjunctions require attention for proper binding (Treisman & Gelade, 1980).

The evidence is scant that features such as color and form, encoded in relatively specialized areas, are functionally bound before spatial awareness. But features such as lines and angles that define an object can be. Although there is preliminary and controversial evidence that surface features may be implicitly bound when explicit spatial awareness is compromised (Wojciulik & Kanwisher, 1998), implicit binding in normal perceivers has not been supported (e.g., Lavie, 1997). If features are bound together implicitly, it is puzzling why Balints patients experience illusory conjunctions, or for that matter, why illusory conjunctions ever occur in normal perceivers. Both clinical and experimental observations with patients like RM suggest that explicit spatial awareness is necessary for the perception of correctly bound features such as color and form, to explicitly individuate objects and to bind action to perceived shape.

7

CHAPTER

Space, Brains, and Consciousness

> Space is a necessary representation *a priori* which serves the foundation of all external intuitions. We never can imagine or make a representation to ourselves of the non-existence of space.
>
> —Immanuel Kant (1724–1804)

The human brain is a marvelous machine that defines every individual's reality. Much of that reality relies on a concept of space. I am a distinct "self," partially because within this skin is "me" and outside of it is everyone and everything else. An object is distinct from all others, partially because it is in a different location or can be described in a different spatial frame from other objects. Grounds and surfaces have volume created by boundaries and are spatially arranged relative to other planes that contribute to the global unity of space as we perceive it. Under normal conditions our brains easily compute the spaces in which we live. But the space that emerges into conscious experience appears to require multiple internal representations that begin by encoding specialized information in parallel and are then integrated into the rich space/object hierarchy we see.

Even imaginary creatures require some sort of spatial reference frame that is internally available. Santa Claus lives at the North Pole because he must "be" somewhere, if only in our imaginations. His imaginary existence relies on spatial attributes. He has breadth and height as well as all sorts of distinctive features that are in their proper places. His cheeks are rosy, and his beard is white. There is a frame of reference that "knows" where his parts are located. When the heroine of a novel is described as svelte and lanky, we do not need to be told that she has a torso with her flat stomach facing forward and thin arms projecting from each shoulder. We all have a spatial frame on which to hang her various body parts, and she emerges in consciousness with the proper spatial relationships.

Spatial frames are so fundamental to our everyday lives that we often take them for granted. We assume that we see space correctly (i.e., that external space is isomorphic to our internal representation of it), and yet this is not always the case. The brain normalizes or arranges "objects" to conform to our lay notions of coordinate geometry. Los Angeles gets moved due south of San Francisco, and Portland, Maine gets pushed north of Portland, Oregon (see Tversky & Schiano, 1989 for a full account of effects such as these). Horizontal and vertical axes are perceived more rapidly than oblique ones and have a stronger influence in perception (Palmer & Hemingway, 1978). Faces are in front and tails in back, with left and right falling out accordingly. These are systematic descriptors that can be applied to most visual stimuli. We use them all the time without thinking about it much.

But when space is explicitly denied, as sometimes observed after brain injury, its fundamental importance for perception and awareness becomes painfully obvious. The consequences can be grim, but the type of brain injury that affects spatial abilities has taught us an enormous amount about the role of space in cognition, and perhaps something about the nature of consciousness itself.

Lessons about Consciousness from the Study of Spatial Deficits

The confluence of cognitive and neural sciences has begun to scientifically test philosophical arguments that seemed out of reach not long ago (e.g., Kant's statement that introduces this chapter). Neuropsychological cases have shown that explicit spatial information is not necessary to perceive an object, even when that object is rather complex, but it is necessary to see more than one object. It is also not necessary to see features such as red and blue, but it is necessary to bind the features into the objects we normally see.

If one assumed the existence of a single spatial map, it would come as something of a surprise that the severe loss of space does not affect the ability to perceive an object, even if only one at a time, since objects are structures with parts spatially related to other parts. One question that then arises is how a brain that allows the perceiver no spatial information outside of his or her own body continues to allow objects in awareness. I have argued that one way this could be done is through implicit spatial maps that remain even when both parietal lobes are damaged. These maps are not sufficient for spatial awareness of the location of each part, but are sufficient for perceptually organizing the parts and maintaining spatial constancy of the perceived object itself (Robertson & Treisman, in preparation). Once an object or shape enters awareness, locating one of its parts (e.g., the left line in a rectangle) is also compromised by parietal damage.

Bringing the relative locations of parts of an object into awareness is as difficult as bringing the relative locations of objects into awareness. When spatial awareness is disrupted by parietal damage, selection is mostly performed bottom-up, automatically, and based on the "best" form available (in the Gestalt sense) (Humphreys et al., 2000). Once this form has entered awareness, its features can be reported, but the location of those features remains explicitly unknown. Voluntarily switching attention to another object (in another location) is difficult, if not impossible, and co-locating surface features for binding is compromised. When spatial awareness is intact, attending to parts of an object, selecting a new object, and properly binding features via spatial attention are all possible.

☐ Parietal Function and Consciousness

It may be foolish of me to venture into this area, as consciousness and its relationship to functional anatomy is such a controversial and slippery topic. But it is hard to ignore after seeing so many cases in which conscious awareness of some properties of visual displays disappear while others remain after brain injury. Despite what appears as normal visual processing by primary vision, a person may not *see* what the visual system has implicitly encoded and organized.

It is also hard to ignore the topic of consciousness given the evidence from fMRI and ERP studies showing that ventral activity can be just as robust when subjects are aware of the presence of a stimulus as when they are not (see Driver I Vuilleumier, 2001, although see Kanwisher, 2001, and Deouell, 2002, for a discussion of exceptions). So I am going to step into the labyrinth of this topic and make some guesses about what have been called the "easy questions" of conscious experience, namely, How do we become aware of the existence of a stimulus? I will leave to others the "hard questions" such as why consciousness is necessary at all or how we develop a concept of the self with a consciousness separate from others (see Shear, 1999). Instead I will focus on what it is that we become aware of when spatial access is denied and the role spatial knowledge may play in conscious awareness.

I introduced this book with three descriptions of visual-spacial problems that can arise after brain injury: "when there is no there there," "when only half is there," and "when something is not there but there." These roughly correspond to bilateral occipital-parietal damage, unilateral right hemisphere damage including parietal and (less often) frontal damage, and unilateral right temporal lobe damage, respectively.

Bilateral Parietal Damage: When There Is No There There

The parietal lobe is a large area of cortex containing many subareas, but the most well studied cases of spatial deficits in humans after *bilateral* parietal damage come from patients with Balints syndrome. The cortical areas that have been most often associated with this syndrome are the angular gyrus and adjacent occipital extrastriate cortex as well as the supramarginal gyrus (Figure 7.1) (De Renzi, 1982; Hacaen & deAjuriaguerra, 1954; Rizzo & Vecera, 2002). However, cases such as RM's demonstrate that supramarginal damage is not necessary to produce this syndrome, leaving the dorsal occipital (Area 19) and angular gyrus (Area 39) as the prime suspects (Rafal, 2001).

Whatever specific brain lesions are responsible for this syndrome, their behavioral consequences are nearly a complete loss of spatial information outside one's own body. In the most severe cases, the self and its distinct space are left intact, as is the perception of objects (although only one at

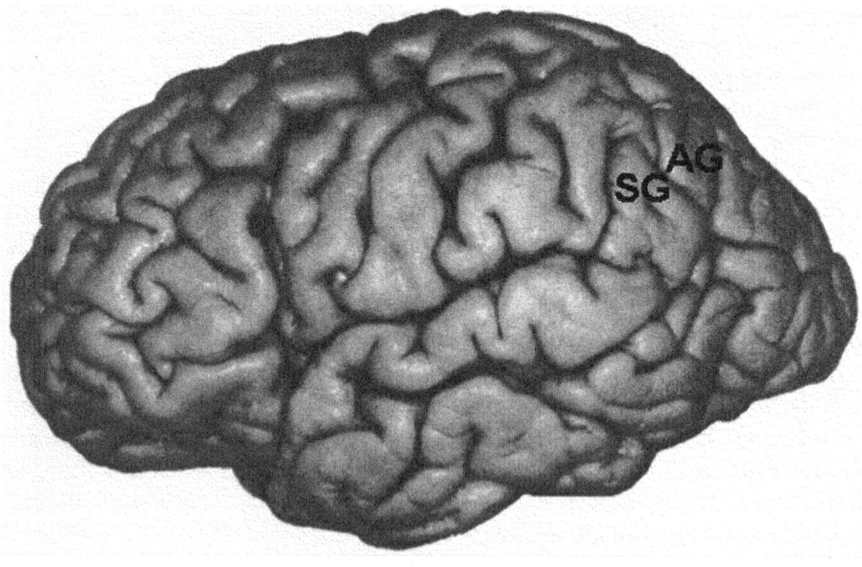

FIGURE 7.1. The angular gyrus (AG) and supramarginal gyrus (SG) of the human brain.

a time), but the *locations* of everything outside the body are lost, as are objects other than the one that enters awareness at any given moment. The following main points derived from studying such patients seem to me to be most relevant for any theory of conscious experience.

- Conscious awareness of one object can exist without conscious knowledge of its location. These patients perceive only one "object" that is presumably spatially organized by the ventral stream of processing. The subthreshold organization of an object seems to follow Gestalt laws of perceptual organization (Humphreys et al., 2000), demonstrating that at least some of these laws operate without explicit access to space. Closure, common fate, connectedness, as well as perceptual learning all contribute to the shapes that define individual objects that have priority for perceptual awareness (see Robertson, 1986). These patients have no volitional control over what they will see next; controlling what enters perceptual awareness through volitional selection is denied. Note that object perception does not disappear when explicit access to space is compromised. Rather, perceptual awareness of more than one object at a time is affected.
- Multiple object properties or surface features that are known to have specialized areas within the primate cortex (e.g., color, size, shape, motion) can be detected without explicit spatial awareness, but correctly binding these features together is compromised when conscious awareness of space is lost. The spatial functions of the inferior parietal lobes are not necessary to consciously perceive features that are present, but their locations and thus their co-locations can become a difficult computational problem. This can result in features being incorrectly bound (or unbound) to form what is consciously perceived.
- Voluntarily switching from local to global frames of reference is severely disturbed, as is switching between one object and another within the same frame. Attentional switches that rely on spatial individuation are all but absent in severe cases of Balints syndrome.
- Not all spatial frames are lost when parietal lobes are damaged. Body frames can remain explicitly intact (Robertson et al., 1997). In other cases without Balints syndrome, environmental frames and body frames can remain intact but with little notion of how to relate the two (Stark et al., 1996).
- Implicit spatial information is clearly present even with one of the most severe explicit spatial deficits seen in neurology. This information supports a fairly accurate organization of multiple spatial frames. The fact that rather complex implicit spatial information does exist may mean that it could become consciously available under the right circumstances. What these circumstances might be is unknown.

- Explicit spatial frames and implicit spatial frames can operate independently, questioning arguments that implicit effects are due to response or decision biases. Implicit spatial performance can remain constant (e.g., spatial Stroop effects), while explicit spatial abilities wax and wane (Robertson et al., 1997).
- Binding perception to action is severely limited when explicit space is lost. Patients with bilateral parietal damage have great difficulty reaching for or looking in the correct direction of the object they do see or forming their hands in a way that spatially conforms to successful grasping (e.g., to pick up a cup).

All these points together first remind us that consciousness itself is not a unitary phenomenon. Even when only addressing the easy question of consciousness, it turns out not to be an all-or-none thing and not to be easy at all. Awareness of a feature such as red in a display can remain intact, but its location can disappear from conscious awareness altogether. Patients with spatial deficits demonstrate that awareness of one property of a stimulus (color) can remain without awareness of another (location) even while the stimulus is in full view and the patient is staring directly at it. They do not appear to be able to explicitly access spatial frames.

Does this mean that perceptual awareness of bound objects and entire scenes is simply the selection of a proper spatial frame? I think not. Although the visual primitives that contribute to the formation of an object's shape seem to be bound without spatial awareness, the evidence for unconscious or preattentive binding of surface features such as color, texture, and size has not received a great deal of support (see discussion in chapter 6). Rather, proper binding of surface features seems to involve an explicit spatial map.[5] But whether implicit feature binding takes place or not, it is clear that features such as color, size, and shape require an explicit spatial map to be bound normally in conscious awareness. For some types of conscious experience, ventral stream processing may be sufficient, but for others it seems to require ventral-parietal interactions.

Unilateral (Typically Right) Parietal Damage: When Only Half Is There

Much of what we know from neuropsychology about perceptual awareness has come from the study of patients with unilateral neglect or extinction (see Driver & Vuilleumier, 2001). Bilateral damage resulting in testable patients with Balints syndrome is rare, while neglect from a unilateral lesion is more common. The fashion has often been to collapse the findings from unilateral and bilateral damage and to emphasize functional similarities. On the surface, Balints syndrome can seem like double unilateral neglect. Awareness of half of space is affected with unilateral damage, while awareness of both sides is affected with bilateral damage; attention to one half of space is spared with unilateral damage, but neither side is spared with bilateral damage; implicit effects are present in both, features are easily detected but conjunctions are not, and so on.

However, the differences may be more informative than the similarities. Unilateral damage can produce object-based neglect, which is difficult to reconcile with simultanagnosia, one of the hallmark symptoms of bilateral dorsal damage where whole objects are perceived. If bilateral damage were simply double neglect, no object should be seen at all, not even one object at a time. Single-object awareness is intact with bilateral damage, while almost everything outside that one object is neglected (with the exception of basic features).

Another potentially important difference is that explicit spatial awareness of the location of parts within all frames of reference (including the object perceived) is compromised by bilateral damage. With bilateral parietal damage, knowing whether the letters N and O are arranged to form the word "no" or "on" is compromised as well as spatially locating a feature on the left or right side of a perceived shape. With unilateral damage, although half (usually the left half) of the display or object may be missed, locating the left side of what *is* perceived is possible.

The majority of anatomical evidence pertaining to neglect suggests that the right temporal/parietal junction including portions of the supramarginal gyrus, and less often the angular gyrus, are most likely to be damaged (see Bisiach & Vallar, 2000; Heilman, Watson, & Valenstein, 1994; Heilman, Watson, Valenstein, & Damasio, 1983; Vallar, 1998; Valar & Parani, 1986). (Figure 7.2). Some recent evidence has suggested that areas more anterior along the superior temporal gyrus may be critical in producing the syndrome (Karnath, Ferber, & Himmelbach, 2001), but this is rather controversial. Most important for the discussion at hand is that in either case, these areas are not centered in the angular gyrus and adjacent occipital lobe, which are most likely to produce Balints syndrome when bilaterally damaged (Hacaen & deAjuriaguerra, 1954; Rafal, 2001).

Given both the behavioral and anatomical factors that differentiate Balints syndrome from many cases of hemineglect, it seems that the two may reflect deficits in rather different functional mechanisms, depending on the location and extent of the lesion. Could it be that Balints syndrome stems from direct assault on spatial representations, which of course would affect attending to space, while the majority of cases of neglect reflect direct damage to an attentional mechanism?

There could be (and very likely are) cases of neglect that result from direct insult to spatial representations themselves (see Bisiach, 1993). For

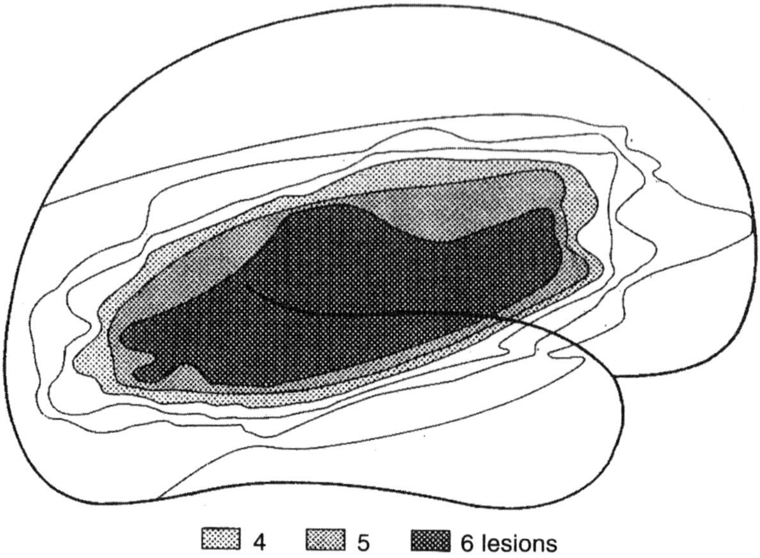

FIGURE 7.2. Area of lesion overlap of six patients with severe unilateral neglect. (Reprinted from Vallar, 1988 with permission of Elsevier Science.)

instance, if unilateral damage were anatomically the same as RM's but on only one side, then half of space might be missing. In this case, features would be detected, as they are by RM, but they could only be located within the part of the spatial map that remains explicitly intact (the ipsilesional side). In fact, this is what occurs with allesthesia (see Deouell, 2002). A stimulus presented on the neglected side is perceived, but it is localized by the patient on their ipsilesional side.

Direct damage to a spatial map might also explain why some patients can enumerate objects that are presented on their neglected side but cannot locate them properly (Vuilleumier & Rafal, 1999). If the contralesional side of a spatial map is dysfunctional, any contralesional items that are perceived would be misplaced to a location that continues to be represented explicitly (the ipsilesional side).

Most patients with visual neglect do not exhibit allesthesia, or at least are not tested for it. But, like all patients with neglect, they miss information on the contralesional side of space. Many patients with neglect seem more aware than others that a left side exists and talk about their problems openly ("I forget to attend to the left"), but they seem to be unable to control the problem volitionally. In these cases, explicit spatial awareness appears intact, but attention is involuntarily biased toward one side.

If the lesion included both the temporal-parietal (more often associated with neglect) and occipital-parietal junction (most often associated with Balints when bilateral), then both the spatial map and attentional mechanisms should be affected and produce profound neglect. In such cases, both the identity and the location of stimuli on the neglected side might not be known unless they were somehow detected by remaining explicit spatial maps in other areas of the brain. These could be accessed by the remaining parietal function of the unaffected hemisphere.

Chronic cases of neglect with relatively small lesions are uncommon, and localizing functional damage can be unreliable in the early days and even months after brain insult, a time when neglect is most prominent. However, the neglect phenomenon itself is variable and the loss of spatial orienting to one side could be disrupted by deficits in spatial representations (perhaps more likely with occipital-parietal damage), spatial attention (more likely with temporal-parietal damage), a combination of both and probably other factors that will be sorted out eventually through rigorous studies.

It is impressive that most patients with neglect as well as patients with Balints syndrome continue to experience pop-out of unique features presented in neglected space. Without extension of the lesion into ventral temporal areas, feature detection should be relatively unaffected, and that is indeed the case, whether unilateral or bilateral damage is present.

I will leave it to the reader to think about other ways in which functional distribution and lesion configuration could predict differences in the constellation of symptoms observed in different patients with unilateral neglect and extinction. I have only listed a few here, and future research will determine whether this line of thinking is correct or not. However, it should be clear that anatomical considerations will prove essential when comparing Balints syndrome to unilateral neglect and relating the findings in each to neurobiological and cognitive correlates of conscious experience.

Ventral (Right) Hemisphere: Not There but There

Studies using functional imaging and electrophysiological procedures with neurological patients as well as studies with normal perceivers have converged in demonstrating a hemispheric difference in processing global and local levels of hierarchically constructed stimuli (e.g., Figure 1.13). Although the exact anatomical locus is debatable, it is clear that global aspects are prioritized by the right hemisphere in posterior ventral areas (either the fusiform, lateral occipital, or temporal lobe with possible limited extension into inferior parietal regions), while local aspects are prioritized by the left. Patients with lesions of the right hemisphere without accompanying neglect may misplace the parts of a figure such that the parts appear jumbled (see the drawing in Figure 1.13). The relative locations of the parts seem to become unglued. That is, the spatial relationship between local elements can be disrupted in a way that produces inaccurate perception of whole objects. This is most clear in patient drawings, but perceptual studies confirm that deficits in processing global shapes are present in perceptual encoding and remain detectable in reaction time measures long after stroke (e.g., Robertson et al., 1988).

Unlike patients with neglect or Balints syndrome, patients who produce scrambled global forms are aware that a global space exists, and given time they can (although not always) identify or describe it accurately. But the elements do not seem to cohere as well or as easily as for normal perceivers. Some investigators have suggested that the problem is one of grouping (Enns & Kingstone, 1995), while others claim that early spatial analysis, tied to spatial frequencies, underlie the phenomenon (Ivry & Robertson, 1998; Sergent, 1982). Whatever the mechanisms turn out to be, conscious awareness of space is not disrupted, only its form is disrupted. The global frame of reference may be altered, producing an incorrectly perceived global object, but awareness of a global space remains. The majority of functional imaging studies have shown inferior temporal activation of the right hemisphere when responding to global information, suggesting a more ventral locus for the effect than the patient data imply.

Since spatial awareness is malformed but not lost in these cases, they may appear as having little to do with space and consciousness. However, the major value for the present purposes is that they contrast with the effects of parietal-occipital damage. Ventral damage does not affect spatial awareness. Patients with bilateral, posterior, ventral lesions may lose the ability to know what an object is, but they do not lose its spatial location (see Alexander & Albert, 1983). Similarly, patients with global processing problems may misplace elements, but some type of explicit spatial map remains even though it may be distorted.

There appear to be certain areas, when damaged, that are more likely to produce deficits in perceptual awareness than others. Parietal damage is especially likely to lead to such deficits. Unilateral neglect may affect attentional exploration of contralesional space so densely that stimuli presented in that part of space disappear from conscious awareness altogether. Neglect may also be affected by damage to the underlying spatial frame and produce a situation where information presented on the left is moved to the right side in perception (allesthesia).

Damage producing Balints syndrome seems to disrupt the computation of an explicit space, which in turn affects spatial attention, object individuation, and binding. It also affects perception of the orientation, reflection, and size of objects that are perceived, all fundamental components of spatial reference frames. Awareness of certain features in a display remains, but the features' locations are unknown and the shapes to which they belong are often misassigned. Damage to more ventral systems can affect the form of a global spatial frame (if in the right hemisphere) resulting in many errors in spatial localization but leaving spatial awareness itself intact.

The evidence together seems to point to the spatial functions of the parietal lobes as being critical for consciousness of a certain kind. Features rise to awareness, as do objects even without explicit spatial knowledge. What is gone is spatial awareness of their locations, orientations, reflections, and sizes, as is awareness of other objects present in the environment as well as the ability to switch between global and local frames of reference.

☐ Spatial Maps and Conscious Perceptions

Losing space does not necessarily lead to a loss of conscious awareness. Rather, losing space leads to losing awareness of multiple objects and makes useless the attentional control that relies on that space. Explicit spatial knowledge is necessary for awareness of some things but not for everything. Kant was right in positing a critical role of spatial knowledge in consciousness, but his thesis was limited by the assumption of space as a unitary entity. If he had had the advantage of the evidence we have today, he might have come to somewhat different conclusions. Spatial maps that support features appear to exist implicitly, but spatial awareness is unnecessary in order to detect these features (Kim & Robertson, 2001). The feature maps may include a spatial frame for encoding, but these frames are not sufficient to consciously know where a feature is located. Implicit spatial maps may also contribute to the formation of objects, but again, they are not sufficient to know explicitly where an object is located. When explicit spatial awareness is all but gone, an object that does enter awareness seems to grab attention (although not to its location), and voluntarily attending to other objects disappears. When one cannot attend to a location, visual awareness seems to become a slave to one object at any given time and to basic features that are present in a display.

The close link between spatial attention and spatial awareness associated with parietal lobes sometimes makes it difficult to determine when attention versus a spatial frame contributes to awareness. When searching for lost keys, we not only move around the house and look here and there, but we use spatial knowledge to scan and decide where we will look next. When explicit space disappears, attentional scanning can no longer be directed by this frame. Under such circumstances, both space-based and object-based attention are affected. Implicitly, objects appear to continue to be formed into spatial hierarchies and appear to conform to the laws of perceptual organization. However, when explicit space is gone, only one of these objects will enter awareness, as will separate features encoded by specialized areas within the ventral cortex.

One might expect that damage to frontal areas of the cortex would more likely result in problems with consciousness, given this area's known involvement in executive function and decision making. Yet *bilateral* frontal damage does not result in the loss of spatial awareness. It does produce memory and judgment problems and can alter personality, but its role in perceiving a stimulus is quite different than that of the parietal lobes (see Chao & Knight, 1995). Like parietal damage, unilateral damage limited to the frontal lobe can produce neglect, but it is less frequent and seems to be of a different sort than unilateral neglect produced by posterior lesions (see Bisiach & Vallar, 2000).

There are other areas of the brain that disrupt conscious experience, but not in such a fundamental way as parietal damage. For instance, anterior temporal lobe lesions can produce a dense amnesia with spared implicit learning. However, conscious awareness of a stimulus on-line is unaffected. Posterior temporal lobe damage can affect object processing, feature encoding, and global or local processing, depending on the side of the lesion, but it does not affect conscious awareness of a stimulus. Likewise, damage to subcortical areas, such as hippocampus, basil ganglia, and amygdala do not generally alter conscious perceptual awareness. Although pulvinar damage can affect spatial awareness, there are such strong connections between this area of the thalamus and parietal lobes that this finding is not surprising.

There seems to be something special about the explicit spatial functions of the parietal system in perceptual awareness. Without this system, perception of the external world is limited to basic features and single objects that pop in and out of view. Yet the objects do not have a location in conscious experience, since there is no explicit spatial frame on which to hang them. Although the spatial frame of the body remains intact, relating items to this frame is also problematic.

Under these circumstances survival would be impossible in natural settings. Features are of little value if their locations are unknown, and objects, even when seen, are difficult to obtain without an accurate spatial map. Although navigation, feeding, and the basic skills of everyday living remain known and can be accurately described by patients with these kinds of spatial problems, the ability to utilize this knowledge is severely compromised without perception of where things are in the environment.

Some Final Comments

It is clear that there are at least two major cortical streams of processing, which on the surface process *what* and *where* or *how*. However, converging evidence demonstrates that there are multiple spatial maps in the brain, many outside the dorsal pathway. There are fine-grained spatial maps in occipital areas, and spatial maps exist in temporal lobes as well, although these may function below the level of awareness. Frontal regions and subcortical areas also contain some rather elaborate spatial maps. However, when both parietal lobes are damaged, none of these is adequate for explicit spatial awareness.

Explicit spatial maps appear to be necessary to switch between hierarchically constructed object/spaces as well as between objects at the same hierarchical level. They are also important for integrating features that are encoded in different specialized cortical areas. It is not clear how explicit spatial maps are computed. They may represent the integration of the multiple spaces that exist in different parts of the brain or they may function to select one or another of these frames for attention. In either case, the dorsal stream interacts extensively with several other areas to support spatial information and correctly bound objects in awareness. These may be spaces that define multiple objects at a particular level or they may be spaces that define objects at different levels of hierarchical structure. They could also be spaces for action, spaces for feature coding, and so on. However many spatial maps there turn out to be, by considering the brain as a space-rich representational device, we may eventually unravel the critical role space plays in conscious awareness, object formation, and integration.

CHAPTER 8

General Conclusions

The main thread weaving through my work over the past 20 years (and thus through the previous chapters) has been an abiding interest in the spatial representations used to reflect what we perceive and believe to be the real world. Space seems to provide the glue that holds our perceptual worlds together so that we can move and act in a manner that is most beneficial for survival. Our sense of vision starts with an analysis of spatially tuned spectral features of the environment but ends with a complete scene that is spatially meaningful. Global spaces are navigated, and the spatial configurations we call objects provide valuable information about where we should attend or look next, what we should avoid, what information we might need for the future, and when to make a decision to act. Space can also be mentally transformed in the mind's eye, dragging its parts (e.g., objects) with it, and it provides a mental workbench for daydreams and problem solving. Without a reasonably accurate representation of space as it exists "out there," we would indeed be left with the "buzzing, booming confusion" that William James suggested we are all born with.

In this book, I have tried to give a glimpse of how the study of spatial representations, and especially the study of spatial deficits, has led to revelations about how mind and brain construct and select spatial maps for perceptual awareness and further analysis. Of course there remains much to learn. I have emphasized neuropsychology, not only because it has been most influential in formulating my own ideas, but also because it has proven useful in building bridges between the psychological and neurobiological. It speaks both languages. Neuropsychological observations also have an uncanny way of provoking insecurities about fundamental assumptions we bring to our experimental designs, the methods we choose, and interpretation of data. They have shown the fallacy of assuming that space is a unitary representation. They have altered interpretations for the role of receptive fields in perception. They have demonstrated that lack of spatial vision can create binding problems. They have articulated when a particular brain area is necessary and when it is not for normal spatial and object perception to occur. The list is long, but these examples are sufficiently representative to demonstrate how neuropsychological investigations can have profound influences on how we formulate the questions that motivate research in the field. They also reveal the limits of relying too heavily on our own perceptual experience of that formation.

☐ Spatial Forms and Spatial Frames

In chapter 1, I described three major ways in which the loss of space affects perception, concluding that objects and space may be dichotomous concepts in everyday language, but they are not so dichotomous for the visual system. Instead, objects and space are reflected in an object/space hierarchy structured within a constellation of spatial frames. The proposal is that for normal perceivers attention selects frames of reference at any of multiple levels to guide further processing. Covert attentional scanning over these frames can then proceed. The frame can define something we call an object, the entire visual field, or various names we have for things in between. This hypothesis seems to capture the deficits observed in Balints syndrome, unilateral neglect and extinction, and even the hemispheric differences seen in certain integrative agnosias (chapter 7). For patients with Balints syndrome, selection of either frames themselves or locations and items within them is virtually absent, at least at an explicit level. For those with unilateral neglect, frame selection appears intact, but one side of the frame is either unattended or missing. For integrative agnosics who favor local parts or objects, selection of global frames is difficult or impossible.

The study of spatial attentional selection is well represented in cognitive neuroscience and has a long and productive history within the cognitive sciences as a whole (Posner & Petersen, 1990). Through collaborations between cognitive scientists and neuroscientists, much has been revealed about the neural systems involved in attracting, shifting, and guiding spatial attention, the subject of chapter 3. However, cognitive neuroscience has been slow to apply cognitive studies of and theoretical approaches to the investigation of neural systems that support perceptual organization, frame selection, and the interactions that occur between these processes and attention. There have been many studies of objects versus space processing and of object-based versus space-based attention, but these concepts are too seldom defined beyond our collective intuition. Yet, in order to select an object, perceptual organizing principals that define a cluster of sensations as belonging together must have already occurred, and part of this process must include how these sensations are spatially related to each other as well as how we decide what cluster is important for the task at hand and how one cluster is integrated into another. All objects appear to be represented in some type of internal spatial frame. Even a circle has a top and bottom in the mind's eye.

Furthermore, in order to select one object among two or more, the two must be identified as two different clusters of sensation (i.e., they require a spatial metric to define where one is relative to the other). The more complex the stimulus becomes, the more perceptual organization is needed before anything like attentional selection can be comprehended. In order for spatial selection to occur, there must be a frame upon which attention can operate. I have argued that this requirement results in multiple spatial frames of reference that define arrangements of parts to each other, objects to each other, groups to each other, as well as their interrelationships within a spatial hierarchy, and ultimately the perception of a unified scene. In this view, the medium for attention is the spatial metric that defines a selected frame.

But what is a frame of reference? If I cannot describe its components, then we are no better off than with naïve notions about what an object or a location in space might be. For this reason, in chapter 2, I adopted the influential approach of Palmer (1999) and described components that are thought to be crucial for frame formation (origin, orientation, sense of direction, and unit size). I then provided neuropsychological evidence showing that these components can be affected independently by lesions in different areas of the human brain. In other words, the data derived from neuropsychological studies demonstrate that the components that define spatial frames of reference do have biological support as being components. This also means that frame components are distributed, leading to a mosaic of ways in which spatial abilities can break down.

How frame formation and frame selection occur and ultimately how attentional and perceptual mechanisms use the results are fertile areas for investigation. For instance, it would be intriguing if the components of relatively global reference frames were represented more strongly by areas within the right hemisphere, while those of relatively local frames were represented within the left. This outcome would be consistent with global/local perceptual deficits observed with damage to posterior areas in the two hemispheres as well as with differences observed with normal perceivers using functional imagery techniques. A very different, but interesting, experiment might be to use evoked potential measures to determine whether covert spatial attentional movements occur after frame selection or in concert with it.

Another fertile area for investigation that arises from the discussions of spatial deficits and descriptions of reference frames is whether brain regions known to be involved in attentional selection are also involved in frame selection. Some initial questions regarding this issue were covered in chapter 3 but many more are ripe for investigation. For instance, are right parietal attentional systems more likely to be activated when scene-based frames are task relevant, but left parietal systems when object-based frames are task relevant? How are these areas involved when switching between frames as opposed to switching between points or items within a frame? Do these areas continue to code their relative locations when frame rotation occurs?

The question of how reference frames at different levels of an object/space hierarchy might interact with attention led me to the discussion of a dichotomy that has became popular within the attention literature: Namely, are there object- and space-based attentional mechanisms (chapters 3 and 4)? The main challenge for the proposition that attention selects objects is how to objectively define an object without reference to space. If we are to seek cognitive and/or neural systems that are object-based, we need to know what an object is a priori, not what it seems to be in our own perceptual experience. Yet this is not often considered seriously in object-based theories of attention except to say that anything that appears as a unit is an object. What appears as a unit to most normal perceivers is an object, but this is not a very satisfying recipe on which to base scientific investigation. Alternatively, if we are to understand cognitive and/or neural systems that are space-based (e.g., spatial attention), we need to be precise about what spatial components we are studying. We also need to be periodically reminded that both objects and space are cognitive creations, not something given automatically by a real world that makes its structures obvious from the start.

I have tried to give examples throughout this book to demonstrate that these are not simply philosophical issues that can be addressed independently of the studies we design. Rather, they bring forth fundamental questions that must be considered when seeking the interpretations of both cognitive and neurobiological evidence that appears on the surface to be tied to objects on the one hand and space on the other. Perhaps by considering frames of reference as the medium for selection, we can avoid the pitfalls of a dichotomy that is fundamentally ambiguous.

Spaces in and out of Awareness

Neuropsychological evidence discussed throughout this book has also demonstrated that the study of space need not be limited to the study of the unitary space we see. The question of how a collection of features becomes a unified object in perception has had a long experimental history, but mechanisms involved in the perceptual unity of space itself have been of lesser interest. It is only recently that we have begun to realize that multiple spatial maps exist, some at a preconscious or implicit level (chapter 5). How the brain integrates, selects, and uses these maps is also important for understanding the perceptual unity of objects and groups of objects.

Both behavioral and neurobiological evidence has shown that many types of stimulus representations can exist below the level of awareness, but space remains essential. We know that everything from basic figure–ground organization to semantics to procedural memory can be represented without our being explicitly aware that anything at all is present (e.g., unilateral neglect). But our scientific studies have been less about the spatial maps than about object features that support these implicit effects. For instance, when performance of a patient with neglect is influenced by semantic associates that he or she has neglected, for some reason it is a common assumption that the stimulus was spatially represented as it usually is but that it simply did not pass some threshold for awareness. So if a drawing of a cat presented in a neglected part of space primes the response to a dog presented in an intact part of space, we most often assume that the brain represented the cat as it normally does except for awareness. But it is not at all clear whether the cat was represented with all of its features, in the orientation in which we presented it, as a spatially extended creature or any of the above. What sort of spatial information represents a drawing of a cat presented below awareness? Likewise, when studies of neglect demonstrate that figure–ground organization can be affected by information in the neglected field, does it mean that the information is represented spatially the same way as when we explicitly perceive the full stimulus? Does implicit spatial information impose a third dimension like its explicit counterpart? Are implicit stimuli hierarchically arranged in multiple spatial frames? Are implicit spaces even Euclidean in the ways that explicit spaces appear to be? The answer to these questions could reveal a great deal about how the sensory world is processed through its various stages before we explicitly perceive a structured world. Implicit spaces may be radial, elongated, two-dimensional, orientation invariant, achromatic, or any number of things, but until we start to explore what the nature of these spaces might be, we can only guess at the properties of implicit representations of the stimuli themselves.

In chapter 5, I allotted a great deal of discussion to studies of rare patients with Balints syndrome who lose explicit spatial awareness but retain spatial information at an implicit level. The data demonstrated that explicit and implicit spatial performance were separable. Variations in explicit spatial awareness did not affect performance on implicit spatial tasks. It is unclear whether implicit spatial information is carried in early visual areas (e.g., subcortical, primary visual cortex), in spaces that control response (e.g., superior colliculus and frontal areas that control eye movements), in any number of other spatial systems, or in all of them. I explored a number of candidates derived from the electrophysiological literature in animals, but this is a topic that is wide open for experimentation. Perhaps future functional imaging studies with humans will reveal what areas of the brain support different types of implicit spatial effects. This may require using imaging procedures with patients who have spatial deficits from isolated brain lesions—a tricky endeavor, but one that is worth pursuing.

☐ The Space That Binds

The binding problem has intrigued cognitive scientists since the demonstration that different features in a display could be bound incorrectly in perception (the red of one item bound to another). Likewise, it has intrigued neuroscientists since the discovery that different features of a display are processed by different specialized areas within the primate cortex. The question is how these separate features are bound together in experience.

As acknowledged in chapter 6, there are many binding problems, and in fact, I have introduced another binding problem in this book: How do multiple spatial frames get bound together to create the unified spatial map we experience? However, the initial question of binding focused on how a surface feature such as color could inappropriately bind to another feature such as shape, as well as what neural mechanisms were involved in binding such features together. In surface feature binding (e.g., binding color to shape), calculating the co-location of features is necessary, while in object binding, the spatial intersections and relationships between parts (e.g., lines, edges, curvature, etc.) are the important variables. In fact, this second type of binding remains relatively intact in Balints patients (at least for one object at a time), while surface feature binding is impaired. These are the same patients for whom implicit spatial maps are present but explicit spatial maps are all but gone. Binding features that are distributed in specialized cortical areas is disrupted when explicit spatial maps are disrupted by damage to posterior "where" systems of the human brain. This led to the argument that spatial functions of the parietal lobes are necessary to co-locate features represented in different feature maps, and thus to bind the features together correctly. It also led to the conclusion that feature binding (but not part binding) occurs in an explicit representation of space. It is through unified spatial maps of the world that accurate feature binding through co-location takes place. As noted in chapter 6 this is a controversial argument but one that is supported by much of the evidence to date.

If this claim is correct, then at an implicit level, the space that supports feature maps may be arguably rather different than the explicit space that guides attention. This conclusion, in turn, has consequences for what form stimuli presented below awareness might take. If the claim is not correct, then it is possible that implicit processing could obtain for everything, including feature binding, with nothing more needed but some act to bring information into awareness.

☐ A Brief Note on Measures

To some readers it might seem somewhat antiquated to put so much emphasis on neuropsychological evidence when we now have methods to observe the normal brain in action. For instance, why not focus more on neuroimaging data to address brain mechanisms of space and object perception?

Neuropsychology can be a messy business. Lesions in humans are difficult to control. They vary in size and location, and even a small lesion affects several functional areas. Nevertheless, it would be difficult to know how the sustained loss of something like spatial awareness affects other functions without examining what happens when that awareness is gone. Furthermore, it is difficult to know what brain areas are necessary for a particular function from functional imagining data alone. An fMRI signal may show activation in one area when doing one task and in another when doing another task, but this does not mean that a lesion in either one of these areas would necessarily disrupt the ability to perform either task. fMRI signal–noise ratios are notoriously poor, and the critical neural activity is very difficult to estimate. In addition, fMRI signals are derived from blood flow, the effect being that some areas are more likely to produce a stronger signal than others, even when all else is equal. Add to this that each trial must collapse over neural activity during at least 2 seconds under the best of conditions (an enormous amount of time for the brain), and the complications of relying on this method alone become quite obvious.

Despite all of these problems, functional imaging has been quite good at defining brain regions that corroborate neuropsychological observations, and it can do so far more precisely. For instance, fMRI studies have articulated areas that are involved in covert (attention) and overt (eye movements) spatial selection, also demonstrating a degree of overlap between the two (Corbetta et al., 1998). The fact that many of these areas, when damaged, disrupt spatial attentional abilities strongly suggests that imaging data are tapping into systems that, at least to some extent, are critical for this ability. This obviously could not be known via neuropsychological evidence alone.

However, when imaging and neuropsychological data conflict, there is a problem. It is then difficult to know where that problem exists, but it seems more likely that imaging has missed something. When a person with damage to C cannot do X but can do Z and a person with damage to D cannot do Z but can do X, it is rather difficult to argue that X and Z tap the same cognitive abilities even if imaging data show strong overlap between C and D.

Along similar lines, when imaging data show that a certain part of the brain is involved in a unique type of processing that has not been suggested before, one still wonders what would happen if we could stop the activity in that area. Would all the other areas that are active during the task be affected, or only some? Would new areas of activation appear? Would nothing at all happen except for a decrease in the signal from the targeted region? To date it is only through damage or temporary deactivation (TMS) that we might know the answers to these questions.

Functional imaging is a wonderful tool, and one I have used on occasion myself, but it must be kept in perspective. Cognitive neuroscience should not be defined by the methods used. The questions are foremost. When data using different methods converge on similar solutions, the results can be extremely powerful (Robertson & Schendel, 2001). But we might need to be reminded periodically that solutions to experience are what we seek (both cognitive and neurobiological). The neuropsychological data have always been an extraordinary part of this endeavor.

I was asked to write this book with an emphasis on my own research, and I have done so. Of course, no research program can stand alone, and I have discussed other supportive, as well as contradictory, evidence from different sources where it seemed best suited. I have made no attempt to be inclusive. That was neither my charge nor my intention. I attempted to stay focused on questions regarding space and objects that have been of most interest to my own research and on some potential answers to what I believe are fundametnal questions. If nothing more, I hope I have succeeded in stimulating thought and future research ideas.

NOTES

Chapter 1

1. Neglect is not typically observed as an abrupt boundary between left and right, and is only discussed this way here for simplicity. Unlike a field cut that affects primary visual centers, the boundary between left and right can be more like a wandering ill-defined and variable diagonal across a display.

Chapter 2

2. One puzzling finding was that only a trend toward a reduction of neglect was found when left flankers were present. Single subject analyses revealed that neglect was significantly reduced for 2 of the 7 patients in this condition. Although the asymmetry in the effects for unilateral left versus right flankers is difficult to interpret, the findings do at least show that the flankers on the neglected side were indeed neglected in visual awareness. This supports the conclusion that neglected flankers were preattentively processed in the bilateral flanker condition and moved the center of attention to the right, but were not effective in moving it with the same strength to the left.

3. In her dissertation Grabowecky demonstrated that these effects were present even when the flankers were different forms with no color in common with targets in the search diamond (e.g., black triangles).

Chapter 5

1. GK has quite different lesion sites than RM (see scan in Humphreys et al., 2000). Whereas RM has bilateral lesions from two separate strokes in the distribution of the middle cerebral artery, GK's lesion on the right appears to be in the distribution of the posterior cerebral artery. His calcarine cortex appears to be infarcted on the right, with the lesion pro-

gressing deeply into white matter that may wll have cut off input into the parietal lobe from the left posterior part of the left hemisphere. This type of lesion typically produces cortical "blindness" in the contralesional field. The lesion on the left included portions of both the temporal and parietal lobes. Nevertheless, GK did show the classical symptoms of Balints syndrome. However, unlike RM, GK is able to accurately perceive spatial orientation of the objects he sees, so some functional differences are present.

Chapter 7

5. Implicit feature binding has received no support from studies with normal perceivers, despite rigorous tests of this hypothesis (Lavie, 1997). Although some evidence for implicit binding with bilateral parietal damage has been reported, it was weakened when other visual features such as brightness contrast were controlled. This remains an issue, and further study is needed to resolve it.

REFERENCES

Alexander, M.P., & Albert, M.L. (1983). The anatomical basis of visual agnosia. In A. Kertesz (Ed.). *Localization in neuropsychology.* New York: Academic Press.

Allport, D.A., Tipper, S.P., Chmiel, N.R.J. (1985). Perceptual integration and postcategorical filtering. In M.I. Posner & O.M. Marin (Eds.), *Attention and Performance, XI.* Hillsdale, NJ: Erlbaum.

Andersen, R.A. (1995). Encoding of intention and spatial location in the posterior parietal cortex. *Cerebral Cortex, 5,* 457–469.

Andersen, R.A., Batista, A.P., Snyder, L.H., Buneo, C.A., & Cohen, Y.E. (2000). Programming to look and reach in posterior parietal cortex. In M. Gazzaniga (Ed.), *The new cognitive neuroscience.* Cambridge, MA: MIT Press.

Andersen, R.A., Essick, G.K., & Seigel, R.M. (1985). Encoding of spatial location by posterior parietal neurons. *Science, 230,* 456–458.

Arguin, M. Cavanagh, P., & Joanette, Y. (1994). Visual feature integration with an attention deficit. *Brain and Cognition, 24,* 44–56.

Asch, S.E., & Witkin, H.A. (1948). Studies in space orientation: Perception of the upright with displaced visual fields. *Journal of Experimental Psychology, 38,* 325–337.

Ashbridge, E., Cowey, A., & Wade, D. (1999). Does parietal cortex contribute to feature binding? *Neuropsychologia, 37,* 999–1004.

Ashbridge, E., Walsh, V., & Cowey, A. (1997). Temporal aspects of visual search studied by transcranial magnetic stimulation. *Neuropsychologia, 35,* 1121–1131.

Ashby, F.G., Prinzmetal, W. , Ivry, R., & Maddox, T. (1996). A formal theory of feature binding in object perception. *Psychological Review, 103,* 165–192.

Baizev, J.S., Ungerleider, L.G., & Desimone, R. (1991). Organization of visual inputs to the inferior temporal and posterior parietal cortex in macaques. *Journal of Neuroscience, 11,* 168–190.

Balint, R. (1909). Seelenlahmung des "Schauens", optische Ataxie, raumliche Storung der Aufmerksamkeit. *Monatshrift fur Psychiatrie und Neurologie, 25,* 5-81. (Translated in *Cognitive Neuropsychology* [1995], *12,* 265–281.

Barnes, L., Beuche, A., & Robertson, L.C. (1999, November). *Inhibition of return in a spatial illusion.* Paper presented at meeting of the Psychonomics Society, Los Angeles.

Bashinski, H.S., & Bacharach, V.R. (1980). Enhancement of perceptual sensitivity as a result of selectively attending to spatial locations. *Perception and Psychophysics, 28,* 241–148.

Baylis, G.C., & Driver, J. (1993). Visual attention and objects; Evidence for hierarchical coding of locations. *Journal of Experimental Psychology: Human Perception and Performance, 19,* 451–470.

Baylis, G.C., Driver, J., Baylis, L.L., & Rafal, R.D. (1994). Reading of letters and words in a patient with Balint's syndrome. *Neuropsychologia, 32,* 1273–1286.

Baylis, G.C., Rafal, R., & Driver, J. (1993). Visual extinction and stimulus repetition. *Journal of Cognitive Neuroscience, 5,* 453–466.

Beck, D.M., Rees, G., Frich, C.D., & Lavie, N. (2001). Neural correlates of change detection and change blindness. *Nature Neuroscience, 4,* 645–650.

Behrmann, M. (2000). Spatial reference frames and hemispatial neglect. In M. Gazzaniga (Ed.), *The new cognitive neuroscience.* Cambridge, MA: MIT Press.

Behrmann, M., & Haimson, C. (1999). The cognitive neuroscience of visual attention. *Current Opinion in Neurobiology, 9,* 158–163.

Behrmann, M., & Moscovitch, M. (1994). Object-centered neglect in patients with unilateral neglect: Effects of left-right coordinates of objects. *Journal of Cognitive Neuroscience, 6,* 1–16.

Behrmann, M., & Tipper, S.P. (1999). Attention accesses multiple reference frames: Evidence from visual neglect. *Journal of Experimental Psychology: Human Perception and Performance, 25,* 83–101.

Benton, A.L., Varney, N.R., & Hamsher, K.D. (1978). Visuo-spatial judgment: A clinical test. *Archives of Neurology,* 35, 364–367.

Berger, A., & Henik, A. (2000). The endogenous modulation of IOR is nasal-temporal asymmetric. *Journal of Cognitive Neuroscience, 12,* 421–428.

Bernstein, L.J., & Robertson, L.C. (1998). Independence between illusory conjunctions of color and motion with shape following bilateral parietal lesions. *Psychological Science, 9,* 167–175.

Berti, A., & Rizzolatti, G. (1992). Visual processing without awareness: Evidence from unilateral neglect. *Journal of Cognitive Neuroscience, 4,* 345–351.

Biederman, I., & Cooper, E.E. (1991). Evidence for complete translational and reflection invariance in visual object priming. *Perception,* 20, 585–593.

Biederman, I., & Gerhardstein, P.C. (1993). Recognizing depth-rotated objects: Evidence and conditions for three-dimensional viewpoint invariance. *Journal of Experimental Psychology: Human Perception & Performance,* 19, 1162–1182.

Bisiach, E. (1993). Mental representation in unilateral neglect and related disorders: The twentieth Bartlett memorial lecture. *Quarterly Journal of Experimental Psychology, 46A,* 435–451.

Bisiach, E., & Luzzatti, C. (1978). Unilateral neglect of representational space. *Cortex, 14,* 129–133.

Bisiach, E., Luzzatti, C., & Perani, D. (1979). Unilateral neglect, representational schema and consciousness. *Brain, 102,* 609–618.

Bisiach, E., Perani, D., Vallar, G., & Berti, A. (1986). Unilateral neglect: Personal and extrapersonal. *Neuropsychologia, 24,* 759–767.

Bisiach, E., & Vallar, G. (2000). Unilateral neglect in humans. In F. Boller, J. Grafman & G. Rizzolatti (Eds.), *Handbook of neuropsychology* (2nd ed.). Amsterdam: Elsevier Press.

Blakemore, C., & Sutton, P. (1969). Side adaptation: A new aftereffect. *Science,* 166, 3902.

Blatt, G.J, Andersen, R.A., & Stoner, G.R. (1990). Visual receptive field organization and cortico-cortical connections of the lateral intraparietal area (area LIP) in macaque. *Journal of Comparative Neurology, 299,* 421–445.

Brain, W.R. (1941). Visual disorientation with special reference to lesions of the right cerebral hemisphere. *Brain, 64,* 244–272.

Braun, J., Koch, C., Lee, K.D., & Itti, L. (2001). Perceptual consequences of multilevel selection. In J. Braun, C. Koch & J.L. Davis (Eds.), *Visual attention and cortical circuits.* Cambridge, MA: MIT Press.

Brooks, J., Wong, T., & Robertson L.C. (2002, April). *Modulation of visual pop-out in hemineglect.* Presented at meeting of the Cognitive Neuroscience Society, San Francisco.

Calvanio, R., Petrone, P.N., & Levine, D.N. (1987). Left visual spatial neglect is both environment-centered and body-centered. *Neurology, 37,* 1179–1183.

Cavada, C., & Goldman-Rakic, P.S. (1989). Posterior parietal cortex in rhesus monkey: Evidence for segregated corticocortical networks linking sensory and limbic areas with the frontal lobe. *Journal of Comparative Neurology, 287,* 422–445.

Chao, L.L., & Knight, R.T. (1995). Human prefrontal lesions increase distractibility to irrelevant sensory inputs. *NeuroReport, 6,* 1605–1610.

Chatterjee, A. (1994). Picturing unilateral spatial neglect: Viewer versus object centered reference frames. *Journal of Neurology, Neurosurgery & Psychiatry, 57,* 1236–1240.

Chatterjee, A., Mennemeier, M., & Heilman, K.M. (1994). The psychophysical power law and unilateral spatial neglect. *Brain and Cognition, 25,* 92–107.

Chelazzi, L. (1999). Serial attention mechanisms in visual search: A critical look at the evidence. *Psychological Research, 62,* 195–219.

Christ, S.E., McCrae, C.S., & Abrams, R.A. (2002). Inhibition of return in static and dynamic displays. *Psychonomic Bulletin & Review, 9,* 80–85.

Cohen, A., & Rafal, R. (1991). Attention and feature integration: Illusory conjunctions in a patient with parietal lobe lesions. *Psychological Science, 2,* 106–110.

Colby, C. (1996). A neurophysiological distinction between attention and intention. In T. Inui & J.L. McClelland (Eds.), *Attention and performance XVI.* Cambridge, MA: MIT Press.

Colby, C.L., & Goldberg, M.E. (1999). Space and attention in parietal cortex. *Annual Review of Neuroscience, 22,* 319–149.

Connor, C.E., Gallant, J.L., Preddie, D.C., & Van Essen, D.C. (1996). Responses in area V4 depend on the spatial relationship between stimulus and attention. *Journal of Neurophysiology, 75,* 1306–1308.

Cooper, A.C.G., & Humphreys, G.W. (2000). Coding space within but not between objects: Evidence from Balint's syndrome. *Neuropsychologia 38,* 723–733.

Cooper, L.A., & Shepard, R.N. (1973). Chronometric studies of the rotation of mental images. In W.C. Chase (Ed.), *Visual information processing.* New York: Academic Press.

Corbetta, M., Akbudak, E., Conturo, T.E., Snyder, A.Z., Ollinger, J.M., Drury, H.A., Linenweber, M.R., Petersen, S.E., Raichle, M.E., Van Essen, D.C., & Shulman, G.L. (1998). A common network of functional areas for attention and eye movements. *Neuron, 21,* 761–73.

Corbetta, M., Shulman, G., Miezin, F., & Petersen, S. (1995). Superior parietal cortex activation during spatial attention shifts and visual feature conjunctions. *Science, 270,* 802–805.

Coren, S., & Hoenig, P. (1972). Effect of non-target stimuli upon length of voluntary saccades. *Perceptual and Motor Skills, 34,* 499–508.

Coslett, H.B., & Saffran, E. (1991). Simultanagnosia: To see but not to see. *Brain, 113,* 1523–1545.

Cowey, A., Small, M., & Ellis, S. (1994). Left visuo-spatial neglect can be worse in far than in near space. *Neuropsychologia, 32,* 1059–1066.

Danziger, S., Fendrich, R., & Rafal, R.D. (1997). Inhibitory tagging of locations in the blind field of hemianopic patients. *Consciousness and Cognition, 6,* 291–307.

Davis, E.T., & Graham, N. (1980). Spatial frequency uncertainty effects in the detection of sinusoidal gratings. *Vision Research, 21,* 705–712.

Delis, D.C., Robertson, L.C., & Efron, R. (1986). Hemisphere specialization of memory for visual hierarchical stimuli, *Neuropsychologia, 24,* 205–214.

Deouell, L.Y. (2002). Prerequisites for conscious awareness: Clues from electrophysiological and behavioral studies of unilateral neglect patients. *Consciousness and Cognition, 11,* 546–567.

Deouell, L.Y., Bentin, S., & Soroker, N. (2000). Electrophysiological evidence for an early (pre-attentive) information processing deficit in patients with right hemisphere damage and unilateral neglect. *Brain, 123,* 353–365.

De Renzi, E. (1982). *Disorders of space exploration and cognition.* Chichester, UK: Wiley.

De Renzi, E., Faglioni, P., & Scotti, G. (1971). Judgment of spatial orientation in patients with focal brain damage. *Journal of Neurology, Neurosurgery & Psychiatry, 34,* 489–495.

De Renzi, E., Faglioni, P., & Villa, P. (1977). Topographical amnesia. *Journal of Neurology, Neurosurgery & Psychiatry, 40,* 498–505.

DeSchepper, B., & Treisman, A. (1996). Visual memory for novel shapes: Implicit coding without attention. *Journal of Experimental Psychology: Learning, Memory & Cognition, 22,* 27–47.

Desimone, R. (2000, November). Keynote address at Society for Neurosciences, New Orleans.

Desimone, R., & Duncan, J. (1995). Neural mechanisms of selective visual attention. *Annual Review of Neuroscience, 18,* 193–222.

Desimone, R., & Shein, S.J. (1987). Visual properties of neurons in area V4 of the macaque: Sensitivity to stimulus form. *Journal of Neurophysiology, 57,* 835–868.

Desimone, R., & Ungerleider, L. (1986). Multiple visual areas in the caudal superior temporal sulcus of the macaque. *Journal of Comparative Neurology, 248,* 164–189.

D'Esposito, M., McGlinchey-Berroth, R., Alexander, M.P., Verfaellie, M., & Milberg, W.P. (1993). Dissociable cognitive and neural mechanisms of unilateral visual neglect. *Neurology, 43,* 2638–2644.

DeValois, R.L., & DeValois, K.K. (1988). *Spatial vision.* New York: Oxford University Press.

DeWeerd, P., Peralta, M.R., Desimone, R., & Ungerleider, L.G. (1999). Loss of attentional stimulus selection after extrastriate cortical lesions in macaques. *Nature: Neuroscience, 2,* 753–758.

Dorris, M.C., Taylor, T.L., Klein, R.M., & Munoz, D.P. (1999). Saccadic reaction times are influenced similarly by previous saccadic metrics and exogenous cuing in monkey. *Journal of Neurophysiology, 81,* 2429–2436.

Downing, C.J. (1988). Expectancy and visual-spatial attention: Effects on perceptual quality. *Journal of Experimental Psychology: Human Perception and Performance, 14,* 188–202.

Drain, M., & Reuter-Lorenz, P.A. (1996). Vertical orienting control: Evidence for attentional bias and "neglect" in the intact brain. *Journal of Experimental Psychology, General, 125,* 139–158.

Driver, J., Baylis, G.C., Goodrich, S.J., & Rafal, R.D. (1994). Axis-based neglect of visual shapes. *Neuropsychologia, 32,* 1353–1365.

Driver, J., Baylis, G.C., & Rafal, R.D. (1992). Preserved figure-ground segregation and symmetry perception in visual neglect. *Nature, 360,* 73–75.

Driver, J., & Halligan, P.W. (1991). Can visual neglect operate in object-centered coordinates? An affirmative single-case study. *Cognitive Neuropsychology, 8,* 475–496.

Driver, J., & Pouget, A. (2000). Object-centered visual neglect or relative egocentric neglect? *Journal of Cognitive Neuroscience, 12,* 542–545.

Driver, J., & Vuilleumier, P. (2001). Perceptual awareness and its loss in unilateral neglect and extinction. *Cognition, 79,* 39–88.

Duhamel, J., Colby, C.L., Goldberg, M.E. (1992). The updating of the representation of visual space in parietal cortex by intended eye movements. *Science, 255,* 90–92.

Duncan, J. (1984). Selective attention and the organization of visual information. *Journal of Experimental Psychology, General, 113,* 501–517.

Duncan, J., & Humphreys, G. (1989). Visual search and stimulus similarity. *Psychological Review, 96,* 433–458.

Edelman, S., & Bulthoff, H.H. (1992). Orientation dependence in the recognition of familiar and novel views of three-dimensional objects. *Vision Research,* 32, 2385–2400.

Efron, R. (1990). *The decline and fall of hemispheric specialization.* Hillsdale, NJ: Erlbaum.

Egeth, H.E., & Yantis, S. (1997). Visual attention: Control, representation and time course. *Annual Review of Psychology, 48,* 269–297.

Eglin, M., Robertson, L.C., & Knight, R.T. (1989). Visual search performance in the neglect syndrome. *Journal of Cognitive Neuroscience, 4,* 372–381.

Eglin, M., Robertson, L.C., Knight, R.T., & Brugger, P. (1994). Search deficits in neglect patients are dependent on size of the visual scene. *Neuropsychology, 4,* 451–463.

Egly, R., Driver, J., & Rafal, R.D. (1994). Shifting visual attention between objects and locations: Evidence for normal and parietal-lesion subjects. *Journal of Experimental Psychology, General, 123,* 161–172.

Egly, R., Rafal, R., Driver, J., & Starreveld, Y. (1994). Hemispheric specialization for object-based attention in a split-brain patients. *Psychological Science, 5,* 380–383.

Egly, R., Robertson, L.C., Rafal, R., & Grabowecky, M. (1995, November). *Implicit processing of unreportable objects in Balint's syndrome.* Poster presented at meeting of the Psychonomic Society, Los Angeles.

Enns, J.R., & Kingstone, A. (1995). Access to global and local properties in visual search for compound stimuli. *Psychological Science, 6,* 283–291.

Epstein, R., & Kanwisher, N. (1998). A cortical representation of the local visual environment. *Nature, 392,* 598–601.

Eriksen, B.A., & Eriksen, C.W. (1974). Effects of noise letters upon the identification of a target letter in a nonsearch task. *Perception and Psychophysics, 16,* 143–149.

Eriksen, C.W., & St. James, J.D. (1986). Visual attention within and around the field of focal attention: A zoom lens model. *Perception and Psychophysics, 40,* 225–240.

Eriksen, C.W. & Yeh, Y (1985). Allocation of attention in the visual field. *Journal of Experimental Psychology: Human Perception and Performance, 11,* 583–597.

Esterman, M., McGlinchey-Berroth, R., & Milberg, W.P. (2000). Parallel and serial search in hemispatial neglect: Evidence for preserved preattentive but impaired attentive processing. *Neuropsychology. 14,* 599–611.

Farah, M.J. (1990). *Visual agnosia.* Cambridge, MA: MIT Press.

Farah, M.J., Brunn, J.L., Wong, A.B., Wallace, M.A., & Carpenter, P.A. (1990). Frames of reference for allocating attention to space: Evidence from the neglect syndrome. *Neuropsychologia, 28,* 335–347.

Filoteo, V.J., Friedrich, F.J., Rabbel, C., & Stricker, J.L. (2002). Visual perception without awareness in a patient with posterior cortical atrophy: Impaired explicit but not implicit processing of global information. *Journal of International Neuropsychological Society, 8,* 461–473.

Filoteo, V.J., Friedrich, F.J., & Stricker, J.L. (2001). Shifting attention to different levels within global-local stimuli: A study of normal participants and a patient with temporal-parietal lobe damage. *Cognitive Neuropsychology, 18,* 227–261.

Fink, G.R., Halligan, P.W., Marshall, J.C., Frith, C.D., Frackowiak, R.S.J., & Dolan, R.J. (1996) Where in the brain does visual attention select the forest and the trees? *Nature, 382,* 626–628.

Fogassi, L., Gallese, V., di Pellegrino, G., Fadiga, L., Gentilucci, M., Matelli, L.G., & Rizzolatti, G. (1992). Space coding by premotor cortex. *Experimental Brain Research, 89,* 686–690.

Franz, E.A. (1997). Spatial coupling in the coordination of complex actions. *The Quarterly Journal of Experimental Psychology, 50A,* 684–704.

Friedman-Hill, S., Robertson, L.C. & Treisman, A. (1995). Parietal contributions to visual feature binding: Evidence from a patient with bilateral lesions. *Science, 269,* 853–855.

Friedman-Hill, Robertson, L.C., Ungerleider, L.G., & Desimone, R. (2000, November). *Impaired attentional filtering in a patient with bilateral parietal lesions.* Paper presented at Society for Neuroscience, New Orleans.

Friedman-Hill, S.R., Robertson, L.C., Ungerleider, L.G., & Desimone, R. (2003). Posterior parietal cortex and the filtering of distractors. *Proceedings of the National Academy of Sciences, 7,* 4263–4268.

Fuentes, L.J., & Humphreys, G.W. (1996). On the processing of "extinguished" stimuli in unilateral visual neglect. An approach using negative priming. *Cognitive Neuropsychology, 13,* 111–136.

Funahashi, S., Bruce, C.J., & Goldman-Rakic, P.S. (1990). Visuospatial coding in primate prefrontal neurons revealed by oculomotor paradigms. *Journal of Neurophysiology, 63,* 814–831.

Funahashi, S., Bruce, C.J., & Goldman-Rakic, P.S. (1993). Dorsolateral prefrontal lesions and oculomotor delayed-response performance: Evidence of mnemonic "scotomas." *Journal of Neuroscience, 13,* 1479–1497.

Gainotti, G., Messerli, P., & Tissor, R. (1972). Qualitative analysis of unilateral spatial neglect in relation to laterality of cerebral lesion. *Journal of Neurology, Neurosurgery and Psychiatry,* 35, 545–550.

Gallant, J.L., Shoup, R.E., & Mazer, J.A. (2000).). A human extrastriate area functionally homologous to macaque V4. *Neuron, 27,* 227–235.

Garner, W.R. (1974). *The processing of information and structure.* New York: Wiley.

Garson, J.W. (2001). (Dis)solving the binding problem. *Philosophical Psychology, 14,* 381–392.

Gattass, R., Sousa, A.P.B., & Covery, E. (1985). Cortical visual areas of the macaque: Possible substrates for pattern recognition mechanisms. In C. Chagas, R. Gattass & C. Gross (Eds.), *Pattern recognition mechanisms.* Vatican City: Pontificia Academia Scientiarum.

Gerstmann, J. (1940). Syndrome of finger agnosia: Disorientation for right and left, agraphia and acalculia. *Archives of Neurology and Psychiatry, 44,* 398–408.

Gianotti, G., Messerli, P., & Tissot, R. (1972). Qualitative analysis of unilateral spatial neglect in relation to laterality of cerebral lesions. *Journal of Neurology, Neurosurgery and Psychiatry, 35,* 545–550.

Gibson, B.S., & Egeth, H. (1994). Inhibition of return to object-based and environment-based locations. *Perception and Psychophysics, 55,* 323–339.

Gilchrist, I.D. Humphreys, G.W., & Riddoch, M.J. (1996). Grouping and extinction: evidence for low-level modulation of visual selection. *Cognitive Neuropsychology, 13,* 1223–1249.

Goldberg, M.E., Colby, C.L., & Duhamel, J.R. (1990). The representation of visuomotor space in the parietal lobe of the monkey. *Cold Spring Harbor Symposia on Quantitative Biology, 55,* 729–739.

Goldman-Rakic, P.S. (1987). Circuitry of primate prefrontal cortex and regulation of behavior by representational memory. In V.B. Mountcastle (Ed.), *Handbook of physiology: The nervous system* (Vol. 5). Bethesda, MD: American Physiological Society.

Goodale, M.A., Milner, A.D., Jakobson, L.S., & Carey, D.P. (1991). A neurological dissociation between perceiving objects and grasping them. *Nature,* 349, 154–156.

Grabowecky, M., Robertson, L.C., & Treisman, A. (1993). Preattentive processes guide visual search: Evidence from patients with unilateral visual neglect. *Journal of Cognitive Neuroscience, 5,* 288–302.

Graham, N. (1981). Psychophysics of spatial-frequency channels. In M. Kubovy & J.R. Pomerantz (Eds.), *Perceptual organization.* Hillsdale, NJ: Erlbaum.

Graham, N., Kramer, P., & Haber, N. (1985). Attending to the spatial frequency and spatial position of near-threshold visual patterns. In M.I. Posner & O.M. Marin (Eds.), *Attention and performance XI.* Hillsdale, NJ: Erlbaum.

Graziano, M.S.A., & Gross, C.G. (1993). A bimodal map of space: Somatosensory receptive field in the macaque putamen with corresponding visual receptive fields. *Experimental Brain Research, 97,* 96–109.

Graziano, M.S.A., & Gross, C.G. (1994). Mapping space with neurons. *Current Directions in Psychological Science, 3,* 164–167.

Graziano, M.S.A., & Gross, C.G. (1998). Spatial maps for the control of movement, *Current Opinion in Neurobiology, 8,* 195–201.

Gross, C.G., & Graziano, M.S.A. (1995). Multiple representations of space in the brain. *The Neuroscientist, 1,* 43–50.

Hacaen, H., & deAjuriaguerra, J. (1954). Balint's syndrome (Psychic paralysis of visual fixation) and its minor forms. *Brain, 77,* 373–400.

Halligan, P.W., & Marshall, J.C. (1991). Left neglect for near but not far space in man. *Nature, 350,* 498–500.

Halligan, P.W., & Marshall, J.C. (1993). Homing in on neglect: A case study of visual search. *Cortex, 29,* 167–174.

Halligan, P.W., & Marshall, J.C. (1997). The art of visual neglect. *Lancet, 350,* 139–140.

Halligan, P.W., & Marshall, J.C. (1998). How long is a piece of string? A study of line bisection in visual neglect. *Cortex, 24,* 321-328.

Han, S.S., Weaver, J.A., Murray, S.O., Kang, X., Yund, E.W., & Woods, D.L. (2002). Hemispheric asymmetry in global/local processing: Effects of stimulus position and spatial frequency. *NeuroImage, 17,* 1290–1299.

Hanaff, M., Michel, F., & Robertson, L.C. (unpublished data, 1996).

Hannay, H.J., Varney, N.R., & Benton, A.L. (1976). Complexity as a determinant of visual field effects for random shapes. *Acta Psychologica, 40,* 29–34.

He, Z.J., & Nakayama, K. (1992). Surfaces versus features in visual search. *Nature 359,* 231–233

Heilman, K.M., Watson, R.T., & Valenstein, E. (1993). Neglect and related disorders. In K.M. Heilman & E. Valenstein (Eds.), *Clinical neuropsychology* (3rd ed.). London: Oxford University Press.

Heilman, K.M., Watson, R.T., & Valenstein, E. (1994). Localization of lesions in neglect and related disorders. In A. Kertesz (Ed.), *Localization and neuroimaging in neuropsychology.* San Diego: Academic Press.

Heilman, K.M., Watson, R.T., Valenstein, E., & Damasio, A.R. (1983). Localization of lesions in neglect. In A. Kertesz (Ed.), *Localization in neuropsychology.* New York: Academic Press.

Heintz, H.J., Hinrichs, H., Scholz, M., Burchert, W., & Mangun, G.R. (1998). Neural mechanisms of global and local processing. A combined PET and ERP study. *Journal of Cognitive Neuroscience, 10,* 485-498.

Hellige, J.B., Cowin, E.L., Eng, T., & Sergent, V. (1991). Perceptual reference frames and visual field asymmetry for verbal processing. *Neuropsychologia, 29,* 929–939

Henderson, J.M., Pollatsek, A., & Rayner, K. (1989). Covert visual attention and extrafoveal information use during object identification. *Perception and Psychophysics, 45,* 196–208.

Henik, A, Rafal, R., & Rhodes, D. (1994). Endogenously generated and visually guided saccades after lesions of the human frontal eye fields. *Journal of Cognitive Neuroscience,* 6, 400–411.

Hillstrom, A.P., & Yantis, S. (1994). Visual motion and attentional capture. *Perception and Psychophysics, 55,* 399–411.

Hof, P.R., Bouras, C., Constantinidis, J., & Morrison, J.H. (1989). Balint's syndrome in Alzheimer's disease: Specific disruption of the occipito-parietal visual pathway. *Brain Research, 493,* 368–375.

Holmes, G. (1919). Disturbances of visual space perception. *British Medical Journal, 2,* 230–233.

Holmes, G., & Horax, G. (1919). Disturbances of spatial orientation and visual attention with loss of stereoscopic vision. *Archives of Neurology and Psychiatry, 1,* 385–407.

Hubel, D.H., & Weisel, T.N. (1959). Receptive fields of single neurons in the cat's striate cortex. *Journal of Physiology, 148,* 574–591.

Humphreys, G.W. (1983). Reference frames and shape perception. *Cognitive Psychology, 25,* 151–196.

Humphreys, G.W. (2001). A multi-stage account of binding in vision: Neuropsychological evidence. *Visual Cognition, 8,* 381–410.

Humphreys, G.W., Cinel, C., Wolfe, J., Olson, A., & Klempen, N. (2000). Fractionating the binding process: Neuropsychological evidence distinguishing binding of form from binding of surface features. *Vision Research, 40,* 1569–1596.

Humphreys, G.W., & Riddoch, M.J. (1987). *To see but not to see: A case study of visual agnosia.* Hillsdale, NJ: Erlbaum.

Humphreys, G.W., & Riddoch, M.J. (1993). Interactions between object and space systems revealed through neuropsychology. In D.E. Meyer & S. Kornblum (Eds.), *Attention and performance, XIV.* Cambridge, MA: MIT Press.

Humphreys, G.W., & Riddoch, M.J. (1994). Attention to within-object and between-object spatial representations: Multiple sites for visual selection. *Cognitive Neuropsychology, 11,* 207–241.

Humphreys, G.W., Romani, C., Olson, A., Riddoch, M.J., & Duncan, J. (1994). Non-spatial extinction following lesions of the parietal lobe in humans. *Nature, 372,* 357–359.

Ivry, R.B., & Robertson, L.C. (1998). *The two sides of perception.* Cambridge, MA: MIT Press.

Johnson, D.M., & Hafter, E.R. (1980). Uncertain-frequency detection: Cueing and condition of observation. *Perception and Psychophysics, 38,* 143–149.

Jolicoeur, P. (1985). The time to name disoriented natural objects. *Memory & Cognition, 16,* 243–275.

Jonides, J. (1983). Further toward a model of the mind's eye movement. *Bulletin of the Psychonomic Society, 21,* 247–250.

Jordan, H., & Tipper, S.P. (1998). Object-based inhibition of return in static displays. *Psychonomic Bulletin & Review, 5,* 504–509.

Jordan, H., & Tipper, S. P. (1999). Spread of inhibition across an object's surface. *British Journal of Psychology, 90,* 495–507.

Kanwisher, N. (2001). Neural events and perceptual awareness. In S. Dehaene (Ed.), *The cognitive neuroscience of consciousness.* Cambridge, MA: MIT Press.

Kanwisher, N., McDermott, J., & Chun, M.M. (1997). The fusiform face area: A module in human extrastriate cortex specialized for face perception. *Journal of Neuroscience, 17,* 4302–4311.

Karnath, H.-O., Christ, K., & Hartje, W. (1993). Decrease of contralateral neglect by neck muscle vibration and spatial orientation of trunk midline. *Brain, 116,* 383–396.

Karnath, H.-O., Ferber, S., & Himmelbach, M. (2001). Spatial awareness is a function of the temporal not the posterior parietal lobe. *Nature, 411,* 950–953.

Karnath, H-O., Ferber, S., Rorden, C., & Driver, J. (2000). The fate of global information in dorsal simultanagnosia. *Neurocase, 6,* 295–306.

Kentridge, R.W., & Heywood, C.A. (2001). Attention and alerting: Cognitive processes spared in blindsight. In B. deGelder, E. deHaan, & C. Heywood (Eds.), *Out of mind: Varieties of unconscious processes.* Oxford: Oxford University Press.

Kim, M.-S., & Cave, K. (1995). Spatial attention in visual search for features and feature conjunctions. *Psychological Science, 6,* 376–380.

Kim, N., Ivry, R., & Robertson, L.C. (1999). Sequential priming in hierarchical figures: Effect of absolute and relative size. *Journal of Experimental Psychology: Human Perception & Performance, 25,* 715–729.

Kim, M.-S., & Robertson, L.C. (2001). Implicit representations of visual space after bilateral parietal damage. *Journal of Cognitive Neuroscience, 13,* 1080–1087.

Kimchi, R., & Palmer, S. (1982). Form and texture in hierarchically constructed patterns. *Journal of Experimental Psychology: Human Perception and Performance, 8,* 521–535.

Kinsbourne, M. (1970). The cerebral basis of lateral asymmetries in attention. *Acta Psychology, 15,* 151–196.

Kinsbourne, M. (1987). Mechanisms of unilateral neglect. In M. Jeannerod (Ed.), *Neuropsychological and physiological aspects of spatial neglect.* Amsterdam: Elsevier.

Klein, R. (1988). Inhibitory tagging system facilitates visual search. *Nature, 334,* 431–431.

Klein, R. (2000) Inhibition of return. *Trends in Cognitive Sciences, 4,* 138–146.

Koch, C., & Crick, F. (1994). Some further ideas regarding the neuronal basis of awareness. In C. Koch & J.L. Davis (Eds.), *Large-scale neuronal theories of the brain.* Cambridge, MA: MIT Press.

Koffka, K. (1935). *Principles of gestalt psychology.* New York: Harcourt & Brace.

Kopferman, H. (1930). Psychologishe Untersuchungen uber die Wirkung Zweidimensionaler korperlicher Gebilde. *Psychologische Forschung, 13,* 293–364.

Kramer, A.F., & Watson, S.E. (1996). Object-based visual selection and the principle of uniform connectedness. In A.F. Kramer, M.G.H. Coles, & G.D. Logan (Eds.), *Converging operations in the study of visual selective attention.* Washington DC: American Psychological Association.

LaBerge, D. (1990). Thalamic and cortical mechanisms of attention suggested by recent positron emission tomographic experiments. *Journal of Cognitive Neuroscience,* 2, 358-372.

LaBerge, D. & Brown, V. (1989). Theory of attentional operation in shape identification. *Psychological Review, 96,* 1010–124.

Ladavas, E. (1987). Is the hemispatial deficit produced by right parietal lobe damage associated with retinal or gravitational coordinates? *Brain, 110,* 167–180.

Ladavas, E., Paladini, R., & Cubelli, R. (1993). Implicit associative priming in a patient with left visual neglect. *Neuropsychologia, 31,* 1307–1320.

Lamb, M.R., Yund, W.E., & Pond, H.M. (1999). Is attentional selection to different levels of hierarchical structure based on spatial frequency? *Journal of Experimental Psychology: General, 128,* 88–94.

Lavie, N. (1997). Feature integration and selective attention: Response competition from unattended distractor features. *Perception & Psychophysics, 59,* 542–556.

Leek, E.C., Reppa, I., & Tipper, S.P. (2003). Inhibition of return for objects and locations in static displays. *Perception & Psychophysics, 65,* 388–395.

Livingstone, M., & Hubel, D. (1988). Segregation of form, color, movement & depth: Anatomy, physiology and perception, *Science, 240,* 740–749.

Logan, G.D. (1995). Linguistic and conceptual control of visual spatial attention. *Cognitive Psychology, 28,* 103–174.

Logan, G.D. (1996). Top-down control of reference frame alignment in directing attention from cue to target. In A.F. Kramer, M.G.H. Coles, & G.D. Logan. (Eds.), *Converging operations in the study of visual selective attention.* Washington, DC: American Psychological Association.

Luck, S.J., & Hillyard, S.A. (1995). The role of attention in feature detection and conjunction discrimination: An electrophysiological analysis. *International Journal of Neuroscience, 80,* 281–297.

Luck, S.J., Massimo, G., McDermott, M.T., & Ford, M.A. (1997). Bridging the gap between monkey neurophyisolgy and human perception: An ambiguity resolution theory of visual selective attention. *Cognitive Psychology, 33,* 64–87.

Luria, A.R. (1959). Disorders of 'simultaneous perception' in a case of bilateral occipito-parietal brain injury. *Brain, 83,* 437–449.

Lynch, L.C., Graybiel, A.M., & Lobeck, L.J. (1985) The differential projection of two cytoarchitectural subregions of the inferior parietal lobule of macaque upon the deep layers of the superior colliculus. *Journal of Comparative Neurology, 235,* 241–254.

Maljkovic, V., & Nakayama, K. (1994). Priming of popout: I. Role of features. *Memory and Cognition, 22,* 657–672.

Mangun, G.R., Heinze, H.J., Scholz, M., & Hinrichs, H. (2000). Neural activity in early visual areas during global and local processing: A reply to Fink, Marshall, Halligan and Dolan. *Journal of Cognitive Neuroscience, 12*, 357–359.

Marr, D. (1982). *Vision*. San Francisco: W.H. Freeman.

Marr, D., & Poggio, T. (1979). A computational theory of human stereo vision. *Proceedings of the Royal Society of London, B, 204*, 301–328.

Marr, D., & Ullman, S. (1981). Directional selectivity and its use in early processing. *Proceedings of the Royal Society of London, B, 211*, 151–180.

Marshall, J.C., & Halligan, P.W. (1989). Does the midsaggital plane play any privileged role in "left neglect." *Cognitive Neuropsychology, 6*, 403–422.

Marshall, J.C., & Halligan, P.W. (1990). Line bisection in a case of visual neglect: Psychophysical studies with implications for theory. *Cognitive Neuropsychology, 7*, 107–130.

Martinez, A., Moses, P., Frank, L., Buxton, R., Wong, E., & Stiles, J. (1997). Hemispheric asymmetries in global and local processing: Evidence form fMRI. *NeuroReport, 8*, 1685–1689.

Mattingley, J.B., Bradshaw, J.L., & Bradshaw, J.A. (1995). The effects of unilateral visuospatial neglect on perception of Muller-Lyer illusory figures. *Perception, 24*, 415–433.

Mattingley, J.B., David, G., & Driver, J. (1997). Pre-attentive filing in of visual surfaces in parietal extinction. *Science, 275*, 671–674.

McCarley, J.S., & He, Z.J. (2001). Sequential priming of 3-D perceptual organization. *Perception and Psychophysics, 63*, 195–208.

McCarthy, R.A., & Warrington, E. K. (1990). *Cognitive Neuropsychology: A Clinical Introduction*. Academic Press: San Diego.

McCloskey, M., & Rapp, B. (2000). Attention-referenced visual representations: Evidence from impaired visual localization. *Journal of Experimental Psychology: Human Perception and Performance, 26*, 917–933.

McCloskey, M., Rapp, B., Yantis, S., Rubin, G., Bacon, W. F., Dagnelie, G., Gordon, B., Aliminosa, D., Boatman, D. F., Badecker, W., Johnson, D. N., Tusa, R. J., & Palmer, E. (1995). A developmental deficit in localizing objects from vision. *Psychological Science, 6*, 112–117.

McGlinchey-Berroth, R., Milberg, W.P., Verfaellie, M., Alexander, M., & Kilduff, P.T. (1993). Semantic processing in the neglected visual field: Evidence from a lexical decision task. *Cognitive Neuropsychology, 10*, 79–108.

Mesulam, M.-M. (1981). A cortical network for directed attention and unilateral neglect. *Annals of Neurology, 4*, 309–325.

Mesulam, M.-M. (1985). Attention, confusional states and neglect. In M.-M. Mesulam, (Ed.), *Principles of behavioral neurology*. Philadelphia: F.A. Davis.

Mesulam, M.-M. (1999). Spatial attention and neglect: Parietal, frontal and cingulate contributions to the mental representation and attentional targeting of salient extrapersonal events. *Philosophical Transcripts of the Royal Society of London, B, 345*, 1325–1346.

Milner, A.D., & Goodale, M.A. (1995). *The visual brain in action*. Oxford: Oxford University Press.

Milner, B. (1965). Visually guided maze learning in man: Effects of bilateral hippocampal, bilateral frontal and unilateral cerebral lesions. *Neuropsychologia, 3*, 317–338.

Mishkin, M., Ungerleider. L.G., & Macko, K.A. (1983). Object vision and spatial vision: Two cortical pathways, *Trends in Neuroscience, 6*, 414-417.

Morais, J., & Bertelson, P. (1975). Spatial position versus ear of entry as determinant of the auditory laterality effects: A stereophonic test. *Journal of Experimental Psychology. Human Perception and Performance, 1*, 253–262.

Moran, J., & Desimone, R. (1985). Selective attention gates visual processing in the extrastriate cortex. *Science, 229*, 782–784.

Moutoussis, K., & Zeki, S. (1997). Functional segretation and temporal hierarchy of the visual perceptive system. *Proceedings of the Royal Society of London, 264,* 1407–1414.

Nadel, L. (1991). The hippocampus and space revisited. *Hippocampus, 1,* 221–229.

Nakayama, K., & Silverman, G.H. (1986). Serial and parallel processing of visual feature conjunctions. *Nature, 320,* 264–265.

Navon, D. (1977). Forest before trees: The precedence of global features in visual perception. *Cognitive Psychology, 9,* 353–383.

Neumann, E., & DeSchepper, B.G. (1992). An inhibition-based fan effect: Evidence for an active suppression mechanism in selective attention. *Canadian Journal of Psychology, 46,* 1–40.

Newcombe, F., & Russell, W.R. (1969). Dissociated visual perceptual and spatial deficits in focal lesions of the right hemisphere. *Journal of Neurology, Neurosurgery & Psychiatry, 32,* 73–81.

O,Craven, K.M., Downing, P.E., & Kanwisher, N. (1999). fMRI evidence for objects as the units of attentional selection, *Nature, 401,* 584–587.

Palmer, S.E. (1980). What makes triangles point? Local and global effects in configurations of ambiguous triangles. *Cognitive Psychology, 12,* 285–305.

Palmer, S.E. (1989). Reference frames in the perception of shape and orientation. In B.E. Shepp & S. Ballesterors (Eds.), *Object perception: Structure and process.* Hilldale, NJ: Erlbaum.

Palmer, S.E. (1999). *Vision sciences.* Cambridge, MA: MIT Press.

Palmer, S.E., & Hemenway, K. (1978). Orientation and symmetry: Effects of multiple, rotational and near symmetries. *Journal of Experimental Psychology: Human Perception & Performance, 4,* 691–702.

Pavlovskaya, M., Glass, I., Soroker, N., Blum, B., & Groswasser, Z. (1997). Coordinate frame for pattern recognition in unilateral spatial neglect. *Journal of Cognitive Neuroscience, 9,* 824–834.

Pavlovskaya, M., Ring, H., Groswasser, Z., & Hochstein, S. (2002). Searching with unilateral neglect. *Journal of Cognitive Neuroscience, 14,* 74–756.

Phan, M.L., Schendel, K.L., Recanzone, G.H., & Robertson, L.R. (2000). Auditory and visual spatial localization deficits following bilateral parietal lobe lesions in a patient with Balint's syndrome. *Journal of Cognitive Neuroscience, 12,* 583–600.

Posner, M.I. (1980). Orienting of attention. *Quarterly Journal of Experimental Psychology, 32,* 3–25.

Posner, M.I., & Cohen, Y. (1984). Components of visual orienting. In H. Bouma & D.G. Bouwhuis (Eds.), *Attention and performance X.* Hillsdale, NJ: Erlbaum.

Posner, M.I., & Petersen, S. (1990). The attention system of the human brain. *Annual Review of Neuroscience, 13,* 25–42.

Posner, M.I., Rafal, R.D., Choate, L., & Vaughn, J. (1985). Inhibition of return: Neural basis and function. *Cognitive Neuropsychology, 2,* 211–228.

Posner, M.I., Snyder, C.R., & Davidson, B.J. (1980). Attention and the detection of signals. *Journal of Experimental Psychology: General, 109,* 160–174.

Posner, M.I., Walker, J.A., Friedrich, F.J., & Rafal, R.D. (1984). Effects of parietal injury on covert orienting of attention. *Journal of Neuroscience, 4,* 1863–1877.

Prinzmetal, W., & Beck, D.M. (2001). The tilt-constancy theory of visual illusions. *Journal of Experimental Psychology: Human Perception and Performance, 27,* 206–217.

Prinzmetal, W., Diedrichson, J., & Ivry, R.B. (2001). Illusory conjunctions are alive and well. *Journal of Experimental Psychology: Human Perception & Performance, 27,* 538–546.

Prinzmetal, W., Presti, D., & Posner, M. (1986). Does attention affect visual feature integration? *Journal of Experimental Psychology: Human Perception and Performance, 12,* 361–369.

Pylyshyn, Z.W., & Storm, R.W. (1988). Tracking multiple independent targets: Evidence for a parallel tracking mechanism. *Spatial Vision, 3,* 179–197.

Quinlan, P.T. (1995). Evidence for the use of scene-based frames of reference in two-dimensional shape recognition. *Spatial Vision, 9,* 101–125.

Rafal, R. (1996). Visual attention: Converging operations from neurology and psychology. In A.F. Kramer, M.G.H. Coles, & G. D. Logan (Eds.), *Converging operations in the study of visual selective attention.* Washington, DC: American Psychological Association.

Rafal, R. (1997). Balint syndrome. In T.E. Feinberg & M.J. Farah (Eds.), *Behavioral neurology and neuropsychology.* New York: McGraw Hill.

Rafal, R. (2001). Balint's syndrome. In Behrmann, M. (Ed.), *Disorders of visual behavior, Vol. 4.* Amsterdam: Elsevier Science

Rafal, R.D., Calabresi, P., Brennan, C.W., & Sciolto, R.K. (1989). Saccade preparation inhibits reorienting to recently attended locations. *Journal of Experimental Psychology: Human Perception and Performance, 15,* 673–685.

Rafal, R., Posner, M.I., Freidman, J.H., Inhoff, A.W., & Bernstein, E. (1988). Orienting of visual attention in progressive supranuclear palsy. *Brain, 111,* 267–280.

Rafal, R., & Robertson, L.C. (1995). The neurology of visual attention. In M. Gazzaniga (Ed.), *The Ccognitive neurosciences* (pp. 625–648). Cambridge, MA: MIT Press.

Ratcliff, G., & Newcombe, F. (1973). Spatial orientation in man: Effects of left, right and bilateral posterior cerebral lesions. *Journal of Neurology, Neurosurgery & Psychiatry, 36,* 448–454.

Rayner, K., McConkie, G.W., & Zola, D. (1980). Integrating information across eye movements. *Cognitive Psychology, 12,* 206–226.

Rensink, R.A., O'Regan, J.K., & Clark, J.J. (1997). To see or not to see: the need for attention to perceive changes in scenes. *Psychological Science, 8,* 368–373.

Rhodes, D.L., & Robertson, L.C. (2002). Visual field asymmetries and attention in scenes. *Brain & Cognition, 50,* 95–115.

Rizzo, M., & Vecera, S.P. (2002). Psychoanatomical substrates of Balint's syndrome. *Journal of Neurology, Neurosurgery & Psychiatry, 72,* 162–178.

Rizzolatti, G., Berti, A., & Gallese, V. (2000). Spatial neglect: Neurophysiological bases, cortical circuits and theory. In F. Boller, J. Grafman, & G. Rizzolatti (Eds.), *Handbook of neuropsychology, Volume I* (2nd ed.). Amsterdam: Elsevier Science.

Rizzolatti, G., Gentilucci, M., & Matelli, M. (1985). Selective spatial attention: One center, one circuit or many circuits? In M.I. Posner & O.S.M. Marin (Eds.), *Attention and performance XI.* Hillsdale NJ: Erlbaum.

Rizzolatti, G., Riggio, L., & Sheliga, B.M. (1994). Space and selective attention In M.I.C. Umilta & M. Moscovitch (Eds), *Attention and performance XV.* Hillsdale, NJ: Erlbaum.

Rizzolatti, G., Scandolara, C., Matelli, M., & Gentilucci, M. (1981). Afferent properties of periarcuate neurons in macaque monkeys: Visual responses. *Behavior and Brain Research, 2,* 147–163.

Ro, T., & Rafal, R.D. (1999). Components of reflexive visual orienting to moving objects. *Perception and Psychophysics, 61,* 826–836.

Ro, T., Rorden, C., Driver, J., & Rafal, R. (2001). Ipsilesional biases in saccades but not perception after lesions of the human inferior parietal lobule. *Journal of Cognitive Neuroscience, 13,* 920–929.

Robertson, L.C. (1986). From Gestalt to neo-Gestalt. In T. J. Knapp & L. C. Robertson (Eds.), *Approaches to cognition: Contrasts and controversies.* Hillsdale, NJ: Erlbaum.

Robertson, L.C. (1995). Covert orienting biases in scene-based reference frames: Orientation priming and visual field differences. *Journal of Experimental Psychology: Human Perception and Performance, 21,* 707–718.

Robertson, L.C. (1996). Attentional persistence for features of hierarchical patterns. *Journal of Experimental Psychology: General, 125,* 227–249

Robertson, L.C. (1999). Spatial frequencies as a medium for guiding attention: Reply to Lamb, Yund and Pond. *Journal of Experimental Psychology: General, 128,* 95–98.

Robertson, L.C. (2003). Binding, spatial attention and perceptual awareness. *Nature Reviews: Neuroscience, 4,* 93–102.

Robertson, L.C., & Beuche, A. (2000, April). *Attentional shifts within and between objects. Effects of perceptual structure and cueing.* Paper presented a a meeting of the Cognitive Neuroscience Society, San Francisco.

Robertson, L.C., Egly, R., Lamb, M. R., & Kerth, L. (1993). Spatial attention and cueing to global and local levels of hierarchical structure. *Journal of Experimental Psychology: Human Perception and Performance, 19,* 471–487.

Robertson, L.C., & Ivry, R. (2000). Hemispheric asymmetry: Attention to visual and auditory primitives. *Current Directions in Psychological Science, 9,* 59–63.

Robertson, L.C., & Kim, M.S. (1998). Effects of perceived space on spatial attention. *Psychological Science, 10,* 76–79.

Robertson, L.C., & Lamb, M.R. (1988). The role of perceptual reference frames in visual field asymmetries. *Neuropsychologia, 26,* 145–152.

Robertson, L.C., & Lamb, M.R. (1989). Judging the reflection of misoriented patterns in the right and left visual fields. *Neuropsychologia, 27,* 1081–1089.

Robertson, L.C., Lamb, M.R., & Knight, R.T. (1988). Effects of lesions of temporal-parietal junction on perceptual and attentional processing in humans. *Journal of Neuroscience, 8,* 3757–3769.

Robertson, L.C., & Lamb, M.R. (1991). Neuropsychological contributions to theories of part/whole organization. *Cognitive Psychology, 23,* 299–330.

Robertson, L.C., & Rafal, R. (2000). Disorders of visual attention. In M. Gazzaniga (Ed.). *The new cognitive neuroscience.* Cambridge, MA: MIT Press.

Robertson, L.C., & Schendel, K.L. (2001). Methods and converging evidence in neuropsychology. In J. Grafman (Ed.), *Handbook of neuropsychology* (2nd ed.). New York: Elsevier.

Robertson, L.C., & Treisman, A. (in preparation). Attending to space within and between objects: Implications from a patient with Balints syndrome.

Robertson, L.C., Treisman, A., Friedman-Hill, S., & Grabowecky, M. (1997). The interaction of spatial and object pathways: Evidence from Balint's syndrome. *Journal of Cognitive Neuroscience, 9,* 295–317.

Rock, I. (1973). *Orientation and form.* New York: Academic Press.

Rock, I. (1983). *The logic of perception.* Cambridge, MA: MIT Press.

Rock, I. (1984) *Perception.* New York: Scientific American Books, Inc.

Rock, I. (1990). The frame of reference. In I. Roch (Ed.), *The legacy of Solomon Asch: Essays in cognition and social psychology.* Hillsdale, NJ: Erlbaum.

Rock, I., & Guttman, D. (1981). The effect of inattention on form perception. *Journal of Experimental Psychology: Human Perception & Performance, 7,* 275–285.

Rolls, E.T., Miyashita, Y., Cahusac, P.M.B., Kesner, R. P., Niki, H., Feigenbaum, J. D., & Bach, L. (1989). Hippocampal neurons in the monkey with activity related to the place in which a stimulus is shown. *Journal of Neuroscience, 9,* 1835–1845.

Salo, R., Robertson L.C., & Nordahl, T. (1996). Normal sustained effects of selective attention are absent in unmedicated patients with schizophrenia. *Psychiatry Research, 62,* 121–130.

Sanocki, R., & Epstein, W. (1997). Priming spatial layout of scenes. *Psychological Science, 8,* 374–378.

Saper, A., Soroker, N., Berger, A., & Henik, A. (1999). "Inhibition of return" in spatial attention: Direct evidence for collicular generation. *Nature: Neuroscience, 2,* 1053–1054.

Scarborough, D.L., Gerard, L., & Cortese, C. (1977). Frequency and repetition effects in lexical memory. *Journal of Experimental Psychology: Human Perception and Performance, 3,* 1–17.

Schendel, K.L. (2001). *Spatial reference frames for visual attention: Evidence from healthy and brain damaged human adults*. Unpublished doctoral dissertation, University of California, Davis.

Schendel, K., & Robertson L.C. (2000, November). *Spatial reference frames in object- based attention*. Paper presented at meeting of Psychonomics Society, New Orleans.

Schendel, K., Fahy, J., & Robertson, L.C. (2002, April). *Stimulus driven influences on attentional selection of spatial reference frames in neglect patients*. Paper presented at meeting of the Cognitive Neuroscience Society, San Francisco.

Schendel, K.L., Robertson, L.C., & Treisman, A. (2001). The role of objects in exogenous attention. *Perception & Psychophysics, 63*, 577–594.

Serences, J.T., Schwarzbach, J., Courtney, S.M., Golay, X., & Yantis, S. (2001, November). *Control of object based attention in human cortex*. Presented at meeting of the Society for Neurosciece, San Diego.

Sergent, J. (1982). The cerebral balance of power: Confrontation or cooperation? *Journal of Experimental Psychology: Human Perception and Performance, 8*, 253–272.

Shalev, L., & Humphreys, G.W. (2002). Implicit location encoding via stored representations of familiar objects; Neuropsychological evidence. *Cognitive Neuropsychology, 19*, 721–744.

Shear, J. (Ed.). (1999). *Explaining consciousness: The hard problem*. Cambridge, MA: MIT Press.

Shomstein, S., & Yantis, S. (2002). Object-based attention: Sensory modulation or priority setting? *Perception and Psychophysics, 64*, 41–51.

Singer, W., & Gray, C.M. (1995). Visual feature integration and the temporal correlation hypothesis. *Annual Review of Neuroscience, 18*, 555–586.

Snyder, L.H., Grieve, K.L., Brotchie, P., & Andersen, R.A. (1998). Separate body- and world-referenced representations of visual space in parietal cortex. *Nature, 397*, 887–891).

Sprague, J.M. (1966). Interaction of cortex and superior colliculus in mediation of peripherally summoned behavior in the cat. *Science, 153*, 1544–1547.

Stark, M., Coslett, H.B., & Saffran, E.M. (1996). Impairment of an egocentric map of locations: Implications for perception and action. *Cognitive Neuropsychology, 13*, 481–523.

Stroop, J.R. (1935). Studies of interference in serial verbal reactions. *Journal of Experimental Psychology, 18*, 643–662.

Suzuki, W.A., & Amaral, D.G. (1994). The perirhinal and parahippocampal cortices of the monkey: Cortical afferents. *Journal of Comparative Neurology, 350*, 497–533.

Tarr, M.J., & Bulthoff, H.H. (1995). Is human object recognition better described by geon structural descriptions or by multiple views? Comment on Biederman & Gerhardstein. *Journal of Experimental Psychology: Human Perception & Performance, 21*, 1494–1505.

Taylor, T.L., & Klein, R.M. (1998). On the causes and effects of inhibition of return. *Psychonomic Bulletin & Review, 5*, 625–643.

Tipper, S.P. (1985). The negative priming effect: Inhibitory priming by ignored objects. *Quarterly Journal of Experimental Psychology, 37A*, 571–590.

Tipper, S.P., & Behrmann, M. (1996). Object-centered not scene-based visual neglect. *Journal of Experimental Psychology: Human Perception and Performance, 22*, 1261–1278.

Tipper, S.P., Driver, J., & Weaver, B. (1991). Object-centered inhibition of return of visual attention. *Quarterly Journal of Experimental Psychology, 43A*, 289–298.

Tipper S.P., & Weaver, B. (1998). The medium of attention: Location-based, object-centered or scene-based? In R.D. Wright (Ed.). *Visual Attention*. New York: Oxford University Press.

Tipper. S.P., Weaver B., Jerreat, L.M., & Burak, A.L. (1994). Object-based and environment-based inhibition of return of visual attention. *Journal of Experimental Psychology: Human Perception and Performance, 20*, 478–499.

Treisman, A.M. (1988). Features and objects. *Quarterly Journal of Experimental Psychology, 40*, 201–237.

Treisman, A.M. (1996). The binding problem: *Current Opinion in Neurobiology, 6*, 171–178.

Treisman, A. (1998) Feature binding, attention and object perception. *Philosophical Transactions of the Royal Society, Series B, 353,* 1295–1306.

Treisman,, A.M. (1999). Solutions to the binding problem: Progress through controversy and convergence, *Neuron, 24,* 105–110.

Treisman, A. (in press). Consciousness and perceptual binding. In A. Cleeremans (Ed.), *The unity of consciousness: Binding, integration and dissociation.* Oxford: Oxford University Press.

Treisman, A.M., & Gelade, G. (1980). A feature-integration theory of attention. *Cognitive Psychology, 12,* 97–136.

Treisman, A.M., & Sato, S. (1990). Conjunction search revisited. *Journal of Experimental Psychology: Human Perception & Performance, 16,* 459–478.

Treisman, A.M., & Schmidt, H. (1982). Illusory conjunctions in perception of objects. *Cognitive Psychology, 14,* 107–141.

Tversky, B., & Schiano, D. (1989). Perceptual and conceptual factors in distortions in memory for maps and graphs. *Journal of Experimental Psychology: General, 118,* 387–398.

Ungerleider, L.G., & Mishkin, M. (1982). Two cortical visual systems. In J. Ingle , M.A. Goodale & R.J.W Mansfield (Eds.), *Analysis of visual behavior.* Cambridge, MA: MIT Press

Vallar, G., & Parani, D. (1986). The anatomy of unilateral neglect after right hemisphere stroke lesions: A clinical/CT scan correlation study in man. *Neuropsychologia, 24,* 609–622.

Vallar, G. (1998). Spatial hemineglect in humans. *Trends in Cognitive Sciences, 2,* 87–97.

Vecara, S.P., & Farah, M.J. (1994). Does visual attention select objects or locations? *Journal of Experimental Psychology: General, 123,* 146–160.

Verfaellie, M., Rapcsak, S.Z., & Heilman, K.M. (1990). Impaired shifting of attention in Balint's syndrome. *Brain & Cognition, 12,* 195–204.

Vuilleumier, P., & Rafal, R. (1999). Both means more than two: Localizing and counting in patients with visuospatial neglect. *Nature Neuroscience, 2,* 783–784.

Walsh, V., & Cowey, A. (1998). Magnetic stimulation studies of visual cognition. *Trends in Cognitive Sciences, 2,* 202–138.

Ward. L.M. (1982). Determinants of attention to local and global features of visual forms. *Journal of Experimental Psychology: Human Perception and Performance, 8,* 562–581.

Ward, R., Danziger, S., Owen, V., & Rafal, R. (2002). Deficits in spatial coding and feature binding following damage to spatiotopic maps in the human pulvinar. *Nature: Neuroscience, 5,* 99–100.

Ward, R., Goodrich, S., & Driver, J. (1994). Grouping reduces visual extinction: Neuropsychological evidence for weight-linkage in visual selection. *Visual Cognition, 1,* 101–129.

Warrington, E.K., & James, M. (1986). Visual object recognition in patients with right hemisphere lesions: Axes of features? *Perception,* 15, 355–366.

Warrington, E.K., & Taylor, A.M. (1978). Two categorical stages of object recognition. *Perception,* 7 695–705.

Watt, R. (1988). *Visual processing: Computational, psychophysical and cognitive research.* Hillsdale, NJ: Erlbaum.

Weiskrantz, L. (1986). *Blindsight: A case study and its implications.* Oxford: Oxford University Press.

Wiser, M. (1981, June). *The role of intrinsic axes in shape recognition.* Paper presented at the 3rd Annual Conference of the Cognitive Science Society, Berkeley, CA. (as referenced in Palmer, 1999).

Wojciulik, E., & Kanwisher, N. (1998). Implicit but not explicit feature binding in a Balint's patient. *Visual Cognition, 5,* 157–181.

Wolfe, J. M. (1994). Guided search 2:0: A revised model of visual search. *Psychonomic Bulletin & Review, 1,* 202–238.

Wolfe, J. M. (1998). Inattentional amnesia. In V. Coltheart (Ed.), *Fleeting memories.* Cambridge, MA: MIT Press.

Yamaguchi, S., Yamagata, S., & Kobayashi, S. (2000). Cerebral asymmetry of the "top-down" allocation of attention to global and local features. *Journal of Neuroscience, 20,* 1–5.

Yantis, S. (1992). Multi-element visual tracking: Attention and perceptual organization. *Cognitive Psychology, 24,* 295–340.

Yantis, S. (1993). Stimulus-driven attentional capture. *Current Directions in Psychological Science, 2,* 156–161.

Yantis, S., & Hillstrom, A.P. (1994). Stimulus-driven attentional capture: Evidence from equiluminant visual objects. *Journal of Experimental Psychology: Human Perception and Performance, 20,* 95–107.

Yantis, S., Schwarzbach, J., Serences, J.T., Carlson, R.L., Steinmetz, M.A., Pekar, J.J., & Courtney, S.M. (2002). Transient neural activity in human parietal cortex during spatial attention shifts. *Nature Neuroscience, 5,* 995–1002.

Yantis, S., & Serences, J.T. (2003). Cortical mechanisms of space-based and object-based attentional control. *Current Opinion in Neurobiology, 13,* 187–193.

Yeshurun, Y., & Carrasco, M. (1999). Spatial attention improves performance in spatial resolution tasks. *Vision Research, 39,* 293–306.

Zeki, S.M. (1978). Functional specialization in the visual cortex of the monkey. *Nature, 274,* 423–428.

INDEX

For Product Safety Concerns and Information please contact our EU representative GPSR@taylorandfrancis.com Taylor & Francis Verlag GmbH, Kaufingerstraße 24, 80331 München, Germany

Batch number: 10726534

Printed by Printforce, the Netherlands